One-Dimensional Nanostructures for PEM Fuel Cell Applications

Hydrogen and Fuel Cells Primers

One-Dimensional Nanostructures for PEM Fuel Cell Applications

Shangfeng Du
University of Birmingham, Birmingham, United Kingdom

Christopher Koenigsmann
Fordham University, Bronx, NY, United States

Shuhui Sun
Institut National de la Recherche Scientifique, Varennes, QC, Canada

Series Editor
Bruno G. Pollet
Department of Energy and Process Engineering
Norwegian University of Science and Technology, Trondheim, Norway

ACADEMIC PRESS
An imprint of Elsevier

Academic Press is an imprint of Elsevier
125 London Wall, London EC2Y 5AS, United Kingdom
525 B Street, Suite 1800, San Diego, CA 92101-4495, United States
50 Hampshire Street, 5th Floor, Cambridge, MA 02139, United States
The Boulevard, Langford Lane, Kidlington, Oxford OX5 1GB, United Kingdom

Notices
Knowledge and best practice in this field are constantly changing. As new research and experience broaden our
understanding, changes in research methods, professional practices, or medical treatment may become
necessary.

Practitioners and researchers must always rely on their own experience and knowledge in evaluating and using
any information, methods, compounds, or experiments described herein. In using such information or methods
they should be mindful of their own safety and the safety of others, including parties for whom they have a
professional responsibility.

To the fullest extent of the law, neither the Publisher nor the authors, contributors, or editors, assume any
liability for any injury and/or damage to persons or property as a matter of products liability, negligence or
otherwise, or from any use or operation of any methods, products, instructions, or ideas contained in the
material herein.

British Library Cataloguing-in-Publication Data
A catalogue record for this book is available from the British Library

Library of Congress Cataloging-in-Publication Data
A catalog record for this book is available from the Library of Congress

ISBN: 978-0-12-811112-3

For Information on all Academic Press publications
visit our website at https://www.elsevier.com/books-and-journals

  **Working together
to grow libraries in
developing countries**

www.elsevier.com • www.bookaid.org

Publisher: Mathew Deans
Acquisition Editor: Christina Gifford
Editorial Project Manager: Charlotte Kent
Production Project Manager: Paul Prasad Chandramohan
Cover Designer: Victoria Pearson

Typeset by MPS Limited, Chennai, India

CONTENTS

LIST OF CONTRIBUTORS

Ian Colliard
Fordham University, Bronx, NY, United States

Shangfeng Du
School of Chemical Engineering, University of Birmingham, Birmingham, United Kingdom

Christopher Koenigsmann
Fordham University, Bronx, NY, United States

Yaxiang Lu
Institute of Physics, Chinese Academy of Sciences, Beijing, China

Shuhui Sun
Center for Énergie Matériaux et Télécommunications, Institut National de la Recherche Scientifique, Montreal, QC, Canada

Gaixia Zhang
Center for Énergie Matériaux et Télécommunications, Institut National de la Recherche Scientifique, Montreal, QC, Canada

Bruno G. Pollet is Full Professor of Renewable Energy in the Department of Energy and Process Engineering at the Norwegian University of Science and Technology (NTNU), Trondheim. His research covers a wide range of areas in Electrochemical Engineering, Electrochemical Energy, and Sono-electrochemistry (the use of Power Ultrasound in Electrochemistry) from the development of novel materials, hydrogen fuel cell to water treatment/disinfection demonstrators and prototypes. He was Professor of Energy Materials and Systems at the University of the Western Cape (South Africa) and R&D Director of the National Hydrogen South Africa (HySA) Systems Competence Centre. He earned a Diploma in Chemistry and Material Sciences from the Université Joseph Fourier (France), a BSc (Hons.) in Applied Chemistry from Coventry University (United Kingdom), and an MSc in Analytical Chemistry from The University of Aberdeen (United Kingdom). He also received his PhD in physical chemistry in the field of Electrochemistry and Sonochemistry at the Coventry University Sonochemistry Centre (United Kingdom).

Shangfeng Du After a first degree in Materials Science and Engineering from the Tsinghua University, China and a PhD degree in Chemical Engineering from the Institute of Process Engineering, Chinese Academy of Sciences, China, in 2005, Shangfeng Du moved to the Max Planck Institute for Metals Research, Germany. After that, he joined the Centre for Fuel Cell and Hydrogen Research (CFCHR) at the University of Birmingham (UoB), United Kingdom, supported by Marie Curie Incoming International Fellowship (IIF) (awarded in 2006). At UoB, he collaborated with Prof. Kevin Kendall FRS, a pioneer in the field of particles and fuel cells (builder of the famous JKR adhesion theory, retired in 2011). In 2009, he was awarded a research fellowship from the Science City Research Alliance (SCRA) through the Higher Education Funding Council for England (HEFCE) Strategic Development Fund and established himself as an independent researcher. In 2015, he was appointed as a lecturer and built the Low Temperature Fuel Cell Research Group as part of the Centre. He has spent 10 years researching electrodes for low-temperature fuel cells and the characterization of nanoparticle behavior for energy and health applications. He is recognized for his expertise in the field of one-dimensional (1D) materials for fuel cell applications, and has introduced the unique category of integrates thin film electrodes from 1D nanostructures for PEMFC application. He is an Editorial Board Member of Scientific Reports, and has authored more than 40 original refereed papers, reviews, book, and book chapters.

Christopher Koenigsmann completed a PhD from Stony Brook University, United States, under the mentorship of Prof. Stanislaus S. Wong. He completed postdoctoral studies in the Department of Chemistry and the Energy Sciences Institute at Yale University, United States, with Prof. Charles A. Schmuttenmaer. Currently, he is an Assistant Professor of Chemistry at Fordham University, United States, and is the Director of the Undergraduate Teaching Assistant and Tutor Program. His research group's interests focus on the synthesis and development of functional nanomaterial architectures that are designed to increase the performance and cost-effectiveness of renewable energy and sensor devices.

Shuhui Sun is a Professor at the Institut National de la Recherche Scientifique (INRS), Center for Energy, Materials, and Telecommunications, Canada. His research is focused on the development of multifunctional nanomaterials for Energy and Environmental applications including PEM fuel cells, Batteries, and Wastewater treatment. He has authored four book chapters and more than 90 peer reviewed articles. These publications have earned him to date more than 4000 citations with H-index 32. He is also listed as inventor on two US patents. He received various honors, including Canada Governor General's Academic Gold Medal (2012), fellow member of Global Young Academy (2017), ECS Toyota Young Investigator Fellow (2017), etc.

CHAPTER 1

Introduction

Shangfeng Du
School of Chemical Engineering, University of Birmingham, Birmingham, United Kingdom

The growing pressures from energy demand and climate change have motivated search for alternative clean and sustainable power generation technologies. The fuel cell, which can directly convert chemical energy (e.g., H_2 and methanol) into electrical power at high energy efficiency with low carbon and pollution emission, has been considering as one of the promising candidates for replacing conventional combustion-based power generators. Of various fuel cell technologies available, proton exchange membrane fuel cells (PEMFCs) have been receiving extensive attention, due to their low operating temperature, easy start-up and shutdown, and flexible power ranges in applications. Despite the initial demonstrated application of PEMFCs, the extensive commercialization has been remarkably hindered by several technological challenges, in particular the low catalytic activity, high cost, poor durability and reliability of fuel cell electrodes. In PEMFCs the power generated is from two electrochemical reactions, namely fuel (e.g., H_2 or hydrocarbon) oxidation reaction at the anode and oxygen reduction reaction (ORR) at the cathode. Electrocatalysts are required to promote the on-going of both electrochemical reactions, and Pt is still considered as one of the best electrocatalysts for PEMFCs up to today. Besides the high cost of Pt catalysts, the slow electrode kinetics and irreversible CO poisoning greatly influence catalyst activities in electrodes, and the harsh operational conditions also cause durability issues for a long-term operation (Gasteiger, Kocha, Sompalli, & Wagner, 2005; Shao, Chang, Dodelet, & Chenitz, 2016). All of these challenges drive researchers to develop lost cost, highly active and robust electrodes to make PEMFCs commercially successful.

The commonly used strategies to reduce the loading amount of Pt in PEMFCs while guaranteeing the performance level include dispersing Pt nanoparticles (NPs) on high surface area carbon support (Pt/C),

One-Dimensional Nanostructures for PEM Fuel Cell Applications. DOI: http://dx.doi.org/10.1016/B978-0-12-811112-3.00001-7

optimizing catalyst size, structure, and morphology, and incorporating transition metals (e.g., Ni and Y) to form Pt alloy or hybrid nanostructures. Although excellent catalytic activities are achieved with these spherical NPs, their high surface energy usually induces severe degradation during PEMFC operation. Meier et al. (2014) summarized five major degradation mechanisms of Pt/C NP catalysts in hydrogen fuel cells, and they are concluded as aggregation, dissolution, Ostwald ripening, carbon support corrosion, and detachment from carbon support. To address these issues, novel nanostructures with excellent stability are urgently required for practical applications.

One-dimensional (1D) nanostructures constrained in two dimensions to less than 100 nm such as nanowire, nanotube, and nanorod represent a promising morphology paradigm that may overcome some inherent drawbacks of zero-dimensional NPs. A schematic summary of various morphologies of 1D nanostructures is shown in Fig. 1.1. 1D nanostructures, due to the relative large scale length, can retain

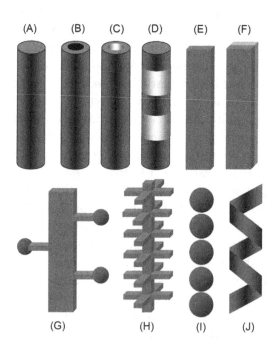

Figure 1.1 A schematic summary of the kinds of quasi-one-dimensional nanostructures. (A) Nanowires, nanowhiskers, and nanorods; (B) core–shell nanostructures; (C) nanotubes; (D) heterostructures; (E) nanobelts; (F) nanotapes; (G) dendrites; (H) hierarchical nanostructures; (I) nanosphere assembly; (J) nanosprings. Reprinted from Kuchibhatla, S. V. N. T., Karakoti, A. S., Bera, D., & Seal, S. (2007). One dimensional nanostructured materials. Progress in Materials Science, 52, 699–913, with permission from Elsevier.

electrical conductivity and also facilitate the electron transport by path-directing effects in catalyst electrodes, so they are less inclined to require a carbon support for remaining dispersing and conducting electrons, thus can potentially address the challenges faced by using the carbon support (Xia, Ng, Wu, Wang, & Lou, 2012; Zhang & Li, 2012). Besides, due to the asymmetric structure, 1D nanostructures are also able to alleviate the dissolution, aggregation, and Oswald ripening that NPs usually suffer from (Koenigsmann & Wong, 2011). However, comparing with 0D nanoparticles, the bulky specific volume of 1D nanostructures decreases their specific surface area, thus a high catalytic activity is not easily achieved. Chapter 2, Advantages and Challenges of One-Dimensional Nanostructures for Fuel Cell Applications, will discuss these features of 1D nanostructures in detail.

Plenty of advanced approaches have been reported to develop 1D nanostructures with well controlled morphologies and compositions in recent years. Xia et al. (2003) classified the fabrication of 1D nanostructures into four categories: (1) anisotropic growth dictated by solid materials with crystallographic structures; (2) anisotropic growth deliberately directed by various templates; (3) anisotropic growth kinetically controlled by appropriate capping agent or supersaturation; (4) miscellaneous methods to yield required 1D nanostructures with intriguing properties. Chapter 3, Preparation of One-Dimensional Catalysts for Fuel Cell Applications, provides a further introduction to all popular preparation approaches of 1D nanostructures for fuel cell applications.

In this primer, based on our very recent review paper published on 1D nanostructures for PEMFC applications (Lu, Du, & Steinberger-Wilckens, 2016), we focus on the very latest efforts on 1D electrocatalysts in recent 5 years, concentrating on the rational design and development of precious metal-based elements, alloys, and hybrid structures, as well as non-precious metal catalysts for two crucial fuel cell reactions, i.e., ORR (see Chapter 4: One-Dimensional Nanostructured Catalysts for Oxygen Reduction Reaction) and the oxidation of hydrocarbon fuels (see Chapter 5: One-Dimensional Nanostructured Catalysts for Hydrocarbon Oxidation Reaction). Considering the requirements and progress for practical applications in PEMFC electrodes (see Chapter 6: Proton Exchange Membrane Fuel Cell Electrodes from One-Dimensional Nanostructures) rather than just pure material research, the challenges facing the development of 1D nanostructures and the prospect to replace

Pt/C NP catalysts for PEMFC applications are explored in Chapter 7, Summary and Perspective.

REFERENCES

Gasteiger, H. A., Kocha, S. S., Sompalli, B., & Wagner, F. T. (2005). Activity benchmarks and requirements for Pt, Pt-alloy, and non-Pt oxygen reduction catalysts for PEMFCs. *Applied Catalysis B-Environmental, 56*, 9−35.

Koenigsmann, C., & Wong, S. S. (2011). One-dimensional noble metal electrocatalysts: a promising structural paradigm for direct methanol fuel cells. *Energy & Environmental Science, 4*, 1161−1176.

Kuchibhatla, S. V. N. T., Karakoti, A. S., Bera, D., & Seal, S. (2007). One dimensional nanostructured materials. *Progress in Materials Science, 52*, 699−913.

Lu, Y. X., Du, S. F., & Steinberger-Wilckens, R. (2016). One-dimensional nanostructured electrocatalysts for polymer electrolyte membrane fuel cells—A review. *Applied Catalysis B-Environmental, 199*, 292−314.

Meier, J. C., Galeano, C., Katsounaros, I., Witte, J., Bongard, H. J., Topalov, A. A., ... Mayrhofer, K. J. J. (2014). Design criteria for stable Pt/C fuel cell catalysts. *Beilstein Journal of Nanotechnology, 5*, 44−67.

Shao, M. H., Chang, Q. W., Dodelet, J. P., & Chenitz, R. (2016). Recent advances in electrocatalysts for oxygen reduction reaction. *Chemical Reviews, 116*, 3594−3657.

Xia, B. Y., Ng, W. T., Wu, H. B., Wang, X., & Lou, X. W. (2012). Self-supported interconnected Pt nanoassemblies as highly stable electrocatalysts for low-temperature fuel cells. *Angewandte Chemie-International Edition, 51*, 7213−7216.

Xia, Y. N., Yang, P. D., Sun, Y. G., Wu, Y. Y., Mayers, B., Gates, B., ... Yan, Y. Q. (2003). One-dimensional nanostructures: Synthesis, characterization, and applications. *Advanced Materials, 15*, 353−389.

Zhang, J. T., & Li, C. M. (2012). Nanoporous metals: fabrication strategies and advanced electrochemical applications in catalysis, sensing and energy systems. *Chemical Society Reviews, 41*, 7016−7031.

CHAPTER 2

Advantages and Challenges of One-Dimensional Nanostructures for Fuel Cell Applications

Gaixia Zhang and Shuhui Sun
Center for Énergie Matériaux et Télécommunications, Institut National de la Recherche Scientifique, Montreal, QC, Canada

Owing to their unique anisotropic morphology the one-dimensional (1D) electrocatalysts (nanowires, nanotubes, nanorods, etc.) can provide various advantages in terms of improved activity and stability, compared with the commonly used zero-dimensional (0D) nanoparticle catalysts. The better activity for the 1D electrocatalysts could be due to several factors that arise from their advantageous structural anisotropy. Specifically, (1) the change in morphology (1D vs. 0D) because the 1D shape could facilitate the reaction kinetics and improve the O_2 diffusion to catalyst surface (Alia et al., 2010); (2) the preferential exposure of highly active, low energy crystal facets (such as 100 and 111 which are commonly observed in 1D noble metal nanostructures) contributes to an increase in the activity of the nanostructures; (3) the adsorbed OH_{ads} species on the noble metal surface could block the active surface for O_2 adsorption and thus have a negative impact on the oxygen reduction reaction (ORR). 1D noble metal nanostructures bear comparatively smooth atomic surfaces with fewer defect sites, which have the corresponding ability to suppress surface passivation by OH_{ads} groups at operating potentials; (4) in addition the 1D structures reduce the number of embedded electrocatalyst sites in the micropores of the carbon supports relative to those in nanogranular Pt.

On the other hand the 1D nanostructures also contribute to the much enhanced catalyst stability, because the up to hundreds of nanometers lengths make the noble metal less vulnerable to dissolution, Ostwald ripening, and aggregation during fuel cell operation compared with the counterpart nanoparticles (Kim, Kim, & Kim, 2009). As for

One-Dimensional Nanostructures for PEM Fuel Cell Applications. DOI: http://dx.doi.org/10.1016/B978-0-12-811112-3.00002-9

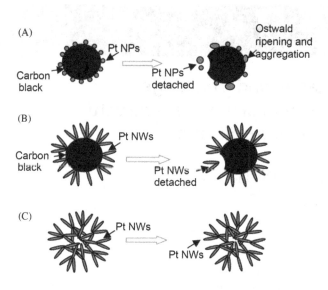

Figure 1.1 Schematic of morphology changes that occur in Pt during accelerated electrochemical cycling. (A) Pt NPs/C (E-TEK); (B) Pt NWs/C; and (C) supportless PtNWs. Reprinted with permission from Sun, S. H., Zhang, G. X., Geng, D. S., Chen, Y. G., Li, R. Y., Cai, M., & Sun, X. (2011). A highly durable Pt nanocatalyst for proton exchange membrane fuel cells: Multiarmed starlike nanowire single crystal. Angewandte Chemie International Edition, 50, 422–426, copyright 2011 Wiley-VCH.

the explanation for the improved durability, as shown in the schematic illustration of morphology changes that occurred in Pt during accelerated electrochemical cycling (Fig. 1.1), the Ostwald ripening and aggregation of Pt, as well as corrosion of the carbon support in the case of the absence of carbon, which are considered on the major contributors to the degradation of fuel cell performance, can be significantly mitigated by introducing 1D Pt morphology (Koenigsmann, Zhou, Adzic, Sutter, & Wong, 2010). Recently, Sun et al. (2011) have continued to address the correlation between structure and electrocatalytic activity by studying the size-dependence of ORR activity in anisotropic systems. Specifically, ultrathin Pt nanowires (diameter of ~4 nm) showed nearly fourfold higher specific activity than that of previously synthesized 200 nm Pt nanotubes and sevenfold higher specific activity than that of commercial Pt nanoparticles. They believe that the surface reconstruction of small diameter nanowires contributes to the enhancement of ORR activity by both decreasing Pt's oxygen affinity and increasing the rate of O–H bond formation. It is also important to highlight that the ultrathin diameter of these nanowires minimizes the quantity of inactive Pt material that is confined within the interior of the wires, thereby lowering their inherent precious metal loading while simultaneously increasing their activity (Sun et al., 2011).

The promising results associated with 1D nanotubes and nanowires clearly demonstrate that the anisotropic character of nanostructured 1D electrocatalysts is uniquely advantageous toward increasing the ORR activity and durability by optimizing the catalyst electronic structure and atomic arrangements at crystal surface. More importantly the results that were achieved without the need for any additional carbon substrate may open a new direction in the fabrication of freestanding self-supported electrodes for hydrogen fuel cells applications, which is a promising step toward reducing the necessary processing and cost of materials.

REFERENCES

Alia, S. M., Zhang, G., Kisailus, D., Li, D. S., Gu, S., Jensen, K., & Yan, Y. S. (2010). Porous Pt nanotubes for oxygen reduction and methanol oxidation reactions. *Advanced Functional Materials, 20*, 3742–3746.

Kim, Y. S., Kim, H. J., & Kim, W. B. (2009). Composited hybrid electrocatalysts of Pt-based nanoparticles and nanowires for low temperature polymer electrolyte fuel cells. *Electrochemistry Communications, 11*, 1026–1029.

Koenigsmann, C., Zhou, W., Adzic, R. R., Sutter, E., & Wong, S. S. (2010). Size-dependent enhancement of electrocatalytic performance in relatively defect-free, processed ultrathin Pt nanowires. *Nano Letters, 10*, 2806–2811.

Sun, S. H., Zhang, G. X., Geng, D. S., Chen, Y. G., Li, R. Y., Cai, M., & Sun, X. (2011). A highly durable Pt nanocatalyst for proton exchange membrane fuel cells: Multiarmed starlike nanowire single crystal. *Angewandte Chemie International Edition, 50*, 422–426.

CHAPTER 3

Preparation of One-Dimensional Catalysts for Fuel Cell Applications

Gaixia Zhang and Shuhui Sun
Center for Énergie Matériaux et Télécommunications, Institut National de la Recherche Scientifique, Montreal, QC, Canada

3.1 NANOWIRES/NANORODS

For the synthesis of nobe metal nanowires and nanorods with well-controlled aspect ratio, several routes have been demonstrated, including the template-assisted approach, the colloidal route using surfactants/capping molecules, and template-free, surfactant-free aqueous solution methods. In the following section, we will focus on some of the recent approaches that have been reported for the synthesis of Pt nanowires and nanorods with well-controlled aspect ratio along with a few shortcomings, which in turn provide important insights into solving the issues related to shape selective growth.

3.1.1 Template-Assisted Approach

Perhaps the most conceptually simple way to generate one-dimensional (1D) nanostructure is to confine their growth with structure-directing templates. Both hard (carbon nanotubes, mesoporous silica, nanoscale channels in polycarbonate (PC), and alumina membranes) and soft (micelles and block copolymers) templates have been successfully used for preparing Pt nanowires or nanorods (Gao, Chen, Peng, & Li, 2002; Govindaraj, Satishkumar, Nath, & Rao, 2000; Liu et al., 2000; Piao, Lim, Chang, Lee, & Kim, 2005; Sakamoto et al., 2004; Song et al., 2007). This route provides several distinct advantages offering a convenient way for producing structurally uniform and often periodically aligned materials in template matrices. For example, Kim and coworkers have demonstrated the preparation of ordered Pt nanowire arrays through an electrochemical deposition route using porous anodic aluminum oxide (AAO) templates (Piao et al., 2005). A typical field-emission

One-Dimensional Nanostructures for PEM Fuel Cell Applications. DOI: http://dx.doi.org/10.1016/B978-0-12-811112-3.00003-0

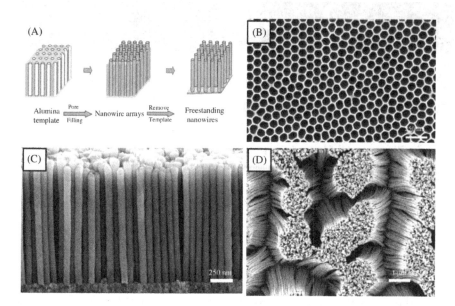

Figure 3.1 (A) Schematic illustration of preparation process of Pt nanowires using porous alumina membrane (PAM) template. FE-SEM images of (B) top view of porous alumina template, and (C) side view and (D) top view of as-prepared Pt nanowire arrays prepared by electrochemical deposition. The diameter of Pt nanowire: ~90 nm. Reprinted from Piao, Y., Lim, H., Chang, J., Lee, W., & Kim, H. (2005). Nanostructured materials prepared by use of ordered porous alumina membranes. *Electrochimica Acta, 50,* 2997–3013, with permission. Copyright 2005 Elsevier.

scanning electron microscopy (FE-SEM) image of a Pt nanowire array is shown in Fig. 3.1, suggesting Pt nanowires to be with uniform diameter and length of ca. 60 nm and 2.6 μm, respectively and high density. Despite the versatility of template-based synthesis, they are restricted by several drawbacks, including the requirement of template removal to obtain a pure product, the limited scope of morphological variation, and the polycrystallinity often associated with the product. In general, template-based methods cannot be easily scaled up to produce nanostructures quickly and cheaply for commercial applications.

Song et al. (2007) synthesized polycrystalline Pt nanowire networks with uniform wire diameters by chemical reduction of a Pt complex using sodium borohydride in a two-phase water-chloroform system. The network of wormlike micelles formed by cetyltrimethylammonium bromide (CTAB) was as a soft template for Pt nanowire growth. A possible formation mechanism (Fig. 3.2A) based on soft templating by the micellar networks contained in chloroform droplets is proposed. The TEM images in Fig. 3.2B and C show that the as-synthesized Pt nanowires have a diameter of 2.2 nm and form large extended wire networks.

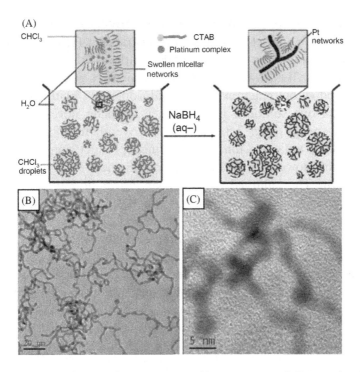

Figure 3.2 (A) Schematic illustration of preparation process of Pt nanowires using soft CTAB template; (B and C) TEM images of the as-synthesized Pt nanowire networks. Reprinted from Song, Y., Garcia, R. M., Dorin, R. M., Wang, H., Qiu, Y., ... Shelnutt, J. A. (2007). Synthesis of Pt nanowire networks using a soft template. *Nano Letters, 7,* 3650–3655, with permission. Copyright 2007 American Chemical Society.

3.1.2 Template-Free Chemical Routes

In the light of the drawback of template-assisted approach, different approaches based on the chemical reduction solution route have been adopted for the synthesis of Pt nanowires and rods. The first attempt to generate single-crystal Pt nanorods or nanowires was accomplished by Chen, Herricks, Geissler, and Xia (2004). Single-crystal Pt nanorods can be obtained by manipulating the reduction kinetics of a polyol process through the addition of a trace amount of iron species (Fe^{2+} or Fe^{3+}) to the reaction system. In a typical polyol process, H_2PtCl_6 is added to and reduced by ethylene glycol in the presence of polyvinyl-pyrrolidone at 110°C. It is worth pointing out that the spherical shape is favored in the context of thermodynamics for an *fcc* metal. By introducing a trace amount of Fe^{2+} or Fe^{3+} species and/or nitrogen at 110 °C the reduction kinetics can be manipulated to obtain Pt nanostructures with different morphologies, including uniform nanorods and branched multipods.

3.1.3 Template-Free, Surfactant-Free Chemical Routes

Recently, Sun, Yang, et al. (2008) demonstrated a facile environmentally friendly route for the large-scale synthesis of single-crystal Pt nanowires and 3D flower-like Pt nanostructures via a very simple chemical reduction reaction of hexachloroplatinic acid with formic acid, in aqueous solution, using neither template nor surfactant or extraneous species. Fig. 3.3A shows the overall morphology of the sample, which indicates that the obtained product consists of large quantities of flower-like structures with diameters in the range of 150−400 nm. The enlarged SEM image shown in Fig. 3.3B reveals that numerous nanowires, with lengths of 100−200 nm, assemble into 3D flower-like superstructures. Fig. 3.3C and D shows the TEM and HRTEM images of Pt nanowires with a diameter of about 4 nm. This work represents a class of new synthetic method with many benefits. First the procedure is very simple and can be performed at room temperature, using commercially available reagents, without the need for complicated templates or potentiostats. Second the process is carried out in an environmentally benign aqueous solution. Third, no surfactant or extraneous species were used in our procedure, which is very

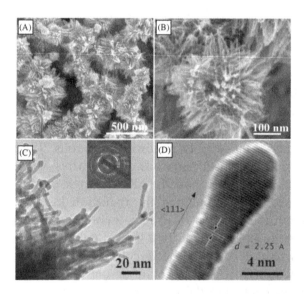

Figure 3.3 (A and B) SEM images of Pt nanoflowers composed of nanowires. (C) TEM image of a Pt nanowires. (D) HRTEM image of the tip of an individual Pt nanowires, indicating it to be a single crystal, with its growth direction along the ⟨111⟩ axis. Modified from Sun, S. H., Yang, D. Q., Villers, D., Zhang, G. X., Sacher, E., & Dodelet, J. P. (2008). Template- and surfactant-free room temperature synthesis of self-assembled 3D Pt nanoflowers from single-crystal nanowires. *Advanced Materials, 20,* 571−574, with permission. Copyright 2008 Wiley-VCH.

important because, as we know, amphiphilic polymers or surfactants typically stabilize high-energy surfaces of nanoparticles. For the realization of the use of Pt nanowires in fuel cells, Sun and coworkers demonstrated the growth of Pt nanowires on different substrates, including the carbon fibers (Sun, Yang, et al., 2008), carbon black spheres (Sun, Jaouen, & Dodelet, 2008), and carbon nanotubes (Sun, Yang, Zhang, Sacher, & Dodelet, 2007), following the same aqueous solution process.

3.2 NANOTUBES

Nanotubes are intriguing to fabricate and study because they possess more advantages than their solid counterparts, such as high-specific surface area and high accessibility of guest species, derived from their inherent hollow space. Up to now the general method employed for the preparation of nanotubes was usually based on template-directed synthesis, e.g., the deposition of metal chemical species on the external surface of solid or supra-molecular core templates such as silver and Te nanowires, or the internal surface of solid or supra-molecular sheath templates such as AAO films or nanoporous PC ones. Templating against existing nanostructures (e.g., wires, rods, belts, and channels) represents a straightforward and powerful route to nanostructures with hollow interiors. The template serves as a scaffold around which a different material is generated in situ and directed to grow into a nanostructure with its morphology complementary to that of the template (Xia et al., 2003). To this end, many different strategies have been successfully demonstrated to synthesis noble metal nanotubes. (1) In one method, it has been shown that the surfaces of nanowires can be directly coated with conformal Pt sheaths to generate coaxial nanocables (He, Law, Fan, Kim, & Yang, 2002). Selective removal of the wires will lead to the formation of Pt nanotubes with well-controlled inner and outer diameters. This route is commonly known as "physical templating." (2) In a second method the nanotubes structures were realized through physical restrictions imposed by the internal surface of hard (PC and alumina membranes) and soft (micelles and block copolymers) templates. This route is commonly known as "nanochannel-directed templating." (3) In a third approach, nanowires (Ag, Cu, and Te) templates react with appropriate Pt salts through a galvanic displacement process which is often referred to as a "template-engaged" process, under carefully controlled conditions to be partially or completely converted into Pt nanotubes without changing the morphology. This route is commonly known as "chemical templating."

3.2.1 Physical Templating

Fig. 3.4A shows a schematic of physical templating. The first step involves the deposition of a desired material on the surface of template. In the second step the template is selectively removed to form nanotubes. Xia and coworkers were the first to use t-Se nanowires with a triangular cross-section as the physical template to synthesize Pt nanotubes (Mayers, Jiang, Sunderland, Cattle, & Xia, 2003). t-Se nanowires' template was synthesized using a chemical or sonochemical method (Mayers, Liu, Sunderland, & Xia, 2003). These nanowires were found to be superior for the deposition of Pt sheaths. The formation of the Pt coating could be achieved by reducing PtCl$_2$ in an alcohol solution. After Pt coating the t-Se template could be easily removed to yield Pt nanotubes using chemical or thermal methods (Mayers, Jiang, et al., 2003). Fig. 3.4B and C shows the typical SEM and TEM image of Pt nanotubes produced using this approach.

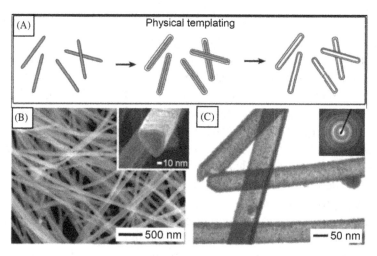

Figure 3.4 (A) A schematic of physical templating; (B) SEM images of Pt nanotubes that were prepared by coating for 18 h, followed by dissolution of Se wires in hydrazine monohydrate. (C) TEM image and electron diffraction pattern of the same sample. Modified from Mayers, B., Jiang, X., Sunderland, D., Cattle, B., & Xia, Y. (2003). Hollow nanostructures of Pt with controllable dimensions can be synthesized by templating against selenium nanowires and colloids. Journal of the American Chemical Society, 125, 13364−13365, with permission. Copyright 2003 American Chemical Society.

3.3 NANOCHANNEL-DIRECTED TEMPLATING

3.3.1 Hard Templates

Electrodeposition in membrane pores is an important method for the synthesis of metal nanotubes. Thus Pt nanotube arrays have been prepared by direct electrodeposition in the nanochannels of AAO templates (Xu et al., 2009). AAO templates with thickness of 6 μm and pore diameters about 250 nm were used to synthesize Pt nanotubes. Pt nanotubes replicate the pore sizes and shapes. The diameter and wall thickness of the Pt nanotubes are about 250 and 45 nm, respectively. After electrodeposition the template was etched away by using 5 wt.% phosphoric acid. The nanotube arrays obtained have a well-controlled microstructure and are polycrystalline with an fcc structure. In addition, electroless deposition has also been used to prepare Pt nanotubes in the nanochannels of AAO template. Electroless deposition is among the most straightforward and efficient methods used to fabricate Pt nanomaterials (Chu, Kawamura, & Mori, 2008; Tsakova, 2008). This method involves the reduction of Pt and possibly other cocatalysts from a metal salt solution onto an electrode surface. The process takes place in acidic, basic, or neutral aqueous solutions and requires a reducing agent. Chu et al. (2008) fabricated Pt nanotubules on aluminum sheets using hydrazine as a reducing agent. Pt nanotubes have also been prepared by the reduction of Pt salt in the pores of track-etched PC membranes that contain 200 nm diameter pores (Zhou, Zhou, Adzic, & Wong, 2009).

3.3.2 Soft Templates

On the basis of the self-organizing structures, various soft templates have been thus used as templating agents for the synthesis of nanostructured materials (Cölfen & Antonietti, 2005; Terrones, Terrones, & Morán-López, 2001; van Bommel, Friggeri, & Shinkai, 2003). The soft templates include those self-assembled and self-organized structures in solution, such as micelle, reverse micelle, microemulsion, and liposome. Surfactant molecules which have hydrophilic heads and lipophilic long carbon chains are commonly used. By using lyotropic mixed-surfactant liquid-crystal templates, Kijima et al. (2004) succeeded in the first synthesis of Pt nanotubes with thin inner diameter and outer diameter by the hydrazine reduction of Pt salts (H_2PtCl_6). In this method, Pt nanotubes are formed by a mixed-surfactant templating mechanism where the hexagonal array of cylindrical rod–like micelles and the aqueous outer shell of these micelles facilitate the growth.

3.3.3 Chemical Templating

Different from physical templating, chemical templating which is based on galvanic replacement provides an amazingly simple and effective method for generating metal nanostructures with hollow interiors. It usually involves some reactions between the template and a precursor to form the desired material. The driving force for the galvanic replacement reaction is the electrical potential difference between two metals, with one metal acting principally as the cathode and the other metal as the anode (Lu, Chen, Skrabalak, & Xia, 2007).

Pt and Pt-based nanotubes with polycrystalline walls and a face-centered cubic structure have been successfully synthesized using silver nanowires as the template (as shown in Fig. 3.5A), because the standard reduction potential of a Pt^{2+}/Pt pair (1.2 V vs. standard hydrogen electrode (SHE)) is higher than that of a Ag^{+}/Ag pair (0.80 V vs. SHE) (Bi & Lu, 2008; Chen, Waje, Li, & Yan, 2007; Sun, Mayers, & Xia, 2002). In this procedure, Ag nanowires are refluxed for 10 min with Pt acetate in aqueous solution. When an aqueous Pt acetate solution is mixed with a dispersion of Ag nanowires, the galvanic replacement reaction generates Pt nanotube (50 nm in diameter, 5−20 μm long, and 4−7 nm wall thickness) whose morphology is complementary to that of the Ag nanowire (Fig. 3.5B).

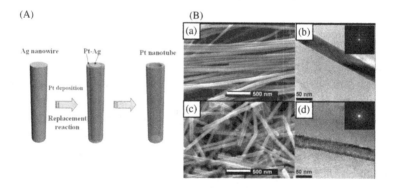

Figure 3.5 (A) Schematic of the fabrication procedure of Pt nanotube arrays via Ag nanowire template. (B) (a) SEM image of AgNWs. (b) TEM image and electron diffraction pattern (inset) of AgNWs. (c) SEM image of PtNTs. (d) TEM image and electron diffraction pattern (inset) of PtNTs. Reprinted from Chen, Z. W., Waje, M., Li, W. Z., & Yan, Y. S. (2007). Supportless Pt and PtPd nanotubes as electrocatalysts for oxygen-reduction reactions. *Angewandte Chemie International Edition, 46*, 4060−4063, with permission. Copyright 2007 Wiley-VCH.

REFERENCES

Bi, Y., & Lu, G. (2008). Facile synthesis of Pt nanofiber/nanotube junction structures at room temperature. *Chemistry of Materials, 20*, 1224−1226.

Chen, J., Herricks, T., Geissler, M., & Xia, Y. (2004). Single-crystal nanowires of Pt can be synthesized by controlling the reaction rate of a polyol process. *Journal of the American Chemical Society, 126*, 10854−10855.

Chen, Z. W., Waje, M., Li, W. Z., & Yan, Y. S. (2007). Supportless Pt and PtPd nanotubes as electrocatalysts for oxygen-reduction reactions. *Angewandte Chemie International Edition, 46*, 4060−4063.

Chu, S. Z., Kawamura, H., & Mori, M. (2008). Ordered integrated arrays of Pd and Pt nanotubules on Al with controllable dimensions and tailored morphologies. *Journal of the Electrochemical Society, 155*, D414−D418.

Cölfen, H., & Antonietti, M. (2005). Mesocrystals: Inorganic superstructures made by highly parallel crystallization and controlled alignment. *Angewandte Chemie International Edition, 44*, 5576−5591.

Gao, T., Chen, Z., Peng, Y., & Li, F. (2002). Fabrication and optical properties of Pt nanowire arrays on anodic aluminium oxide templates. *Chinese Physics, 11*, 1307−1312.

Govindaraj, A., Satishkumar, B. C., Nath, M., & Rao, C. N. R. (2000). Metal nanowires and intercalated metal layers in single-walled carbon nanotube bundles. *Chemistry of Materials, 12*, 202−215.

He, R., Law, M., Fan, R., Kim, F., & Yang, P. (2002). Functional bimorph composite nanotapes. *Nano Letters, 2*, 1109−1112.

Kijima, T., Yoshimura, T., Uota, M., Ikeda, T., Fujikawa, D., Mouri, S., ... Uoyama, S. (2004). Noble-metal nanotubes (Pt, Pd, Ag) from lyotropic mixed-surfactant liquid-crystal templates. *Angewandte Chemie International Edition, 43*, 228−232.

Liu, Z., Sakamoto, Y., Ohsuna, T., Hiraga, K., Terasaki, O., Ko, C. H., ... Ryoo, R. (2000). TEM studies of Pt nanowires fabricated in mesoporous silica MCM-41. *Angewandte Chemie International Edition, 39*, 3107−3110.

Lu, X., Chen, J., Skrabalak, S. E., & Xia, Y. (2007). Galvanic replacement reaction: A simple and powerful route to hollow and porous metal nanostructures. *Proceedings of the Institution of Mechanical Engineers, Part N: Journal of Nanoengineering and Nanosystems, 221*, 1−16.

Mayers, B., Jiang, X., Sunderland, D., Cattle, B., & Xia, Y. (2003). Hollow nanostructures of Pt with controllable dimensions can be synthesized by templating against selenium nanowires and colloids. *Journal of the American Chemical Society, 125*, 13364−13365.

Mayers, B. T., Liu, K., Sunderland, D., & Xia, Y. (2003). Sonochemical synthesis of trigonal selenium nanowires. *Chemistry of Materials, 15*, 3852−3858.

Piao, Y., Lim, H., Chang, J., Lee, W., & Kim, H. (2005). Nanostructured materials prepared by use of ordered porous alumina membranes. *Electrochimica Acta, 50*, 2997−3013.

Sakamoto, Y., Fukuoka, A., Higuchi, T., Shimomura, N., Inagaki, S., & Ichikawa, M. (2004). Synthesis of Pt nanowires in organic−inorganic mesoporous silica templates by photoreduction: formation mechanism and isolation. *The Journal of Physical Chemistry B, 108*, 853−858.

Sun, Y., Mayers, B. T., & Xia, Y. (2002). Template-engaged replacement reaction: a one-step approach to the large-scale synthesis of metal nanostructures with hollow interiors. *Nano Letters, 2*, 481−485.

Sun, S., Jaouen, F., & Dodelet, J. (2008). Controlled growth of Pt nanowires on carbon nanospheres and their enhanced performance as electrocatalysts in PEM fuel cells. *Advanced Materials, 20*, 3900−3904.

Sun, S., Yang, D., Zhang, G., Sacher, E., & Dodelet, J. (2007). Synthesis and characterization of Pt nanowire–carbon nanotube heterostructures. *Chemistry of Materials, 19*, 6376–6378.

Sun, S. H., Yang, D. Q., Villers, D., Zhang, G. X., Sacher, E., & Dodelet, J. P. (2008). Template- and surfactant-free room temperature synthesis of self-assembled 3D Pt nanoflowers from single-crystal nanowires. *Advanced Materials, 20*, 571–574.

Song, Y., Garcia, R. M., Dorin, R. M., Wang, H., Qiu, Y., Coker, E., ... Shelnutt, J. A. (2007). Synthesis of Pt nanowire networks using a soft template. *Nano Letters, 7*, 3650–3655.

Terrones, H., Terrones, M., & Morán-López, J. L. (2001). Curved nanomaterials. *Current Science, 81*, 1011–1029.

Tsakova, V. (2008). How to affect number, size, and location of metal particles deposited in conducting polymer layers. *Journal of Solid State Electrochemistry, 12*, 1421–1434.

van Bommel, K. J. C., Friggeri, A., & Shinkai, S. (2003). Organic templates for the generation of inorganic materials. *Angewandte Chemie International Edition in English, 42*, 980–999.

Xia, Y., Yang, P., Sun, Y., Wu, Y., Mayers, B., Gates, B., ... Yan, H. (2003). One-dimensional nanostructures: Synthesis, characterization, and applications. *Advanced Materials, 15*, 353–389.

Xu, Q. L., Meng, G. W., Wu, X. B., Wei, Q., Kong, M. G., Zhu, X. G., ... Chu, Z. Q. (2009). A generic approach to desired metallic nanowires inside native porous alumina template via redox reaction. *Chemistry of Materials, 21*, 2397–2402.

Zhou, H., Zhou, W., Adzic, R. R., & Wong, S. S. (2009). Enhanced electrocatalytic performance of one-dimensional metal nanowires and arrays generated via an ambient, surfactantless synthesis. *The Journal of Physical Chemistry C, 113*, 5460–5466.

One-Dimensional Nanostructured Catalysts for the Oxygen Reduction Reaction

Ian Colliard and Christopher Koenigsmann
Fordham University, Bronx, NY, United States

Within the cathode half-cell of Proton Exchange Membrane Fuel Cells (PEMFCs), molecular oxygen is reduced to water in a four-electron process (cf. Eq. 4.1) referred to as the oxygen reduction reaction (ORR) (Gewirth & Thorum, 2010; Watanabe, Tryk, Wakisaka, Yano, & Uchida, 2012). The most commonly employed electrocatalysts for ORR consist primarily of platinum and other precious metals (He, Desai, Brown, & Bollepalli, 2005; Li, Lv, Kang, Markovic, & Stamenkovic, 2016; Rabis, Rodriguez, & Schmidt, 2012; Shao, Chang, Dodelet, & Chenitz, 2016). Although Pt is traditionally considered to be a highly effective catalyst for a broad range of reactions, the high cost and low abundance of platinum has driven interest in the development of nanostructured electrocatalysts, which have the advantage of dramatically higher specific surface areas and the potential for a significant reduction in the platinum loading within the cathode (Antolini & Perez, 2011; Chuan-Jian et al., 2010; Mazumder, Lee, & Sun, 2010; Morozan, Jousselme, & Palacin, 2011) Commercially available ORR catalysts consist of platinum nanoparticles physisorbed onto high-surface area carbon supports (Pt NP/C) (Koper, 2009). Despite the gains in specific surface area, the practical use of these catalysts to develop commercially viable PEMFCs has not yet been realized and is largely due to technological shortfalls with the activity and durability of Pt NP/C.

$$O_2 + 4\,H^+ + 4\,e^- \rightarrow 2\,H_2O, \quad E^\circ = 1.23\ V \qquad (4.1)$$

A significant challenge with the use of pure platinum is that it maintains relatively poor ORR activity at potentials close to the thermodynamic potential of 1.23 V (Adzic & Wang, 1998; Adzic & Wang, 2000;

One-Dimensional Nanostructures for PEM Fuel Cell Applications. DOI: http://dx.doi.org/10.1016/B978-0-12-811112-3.00004-2

Gewirth & Thorum, 2010; Wang, Markovic, & Adzic, 2004; Wang, Zhang, & Adzic, 2007). The low activity of platinum requires a significant overpotential of 0.3–0.4 V to achieve reasonable kinetics, which results in a significant reduction in the efficiency of an operating PEMFC. This has led to considerable interest in examining the origin of the overpotential for ORR on Pt. The computational work of Norksøv and coworkers attributes the sluggish kinetics of ORR to the relatively strong adsorption of oxygen species on the surface of Pt (111) (Greeley & Nørskov, 2009; Nørskov et al., 2004; Nørskov, Bligaard, Rossmeisl, & Christensen, 2009). At a low overpotential, the strong binding energy of oxygen adsorbates leads to a high coverage of adsorbed oxide species (i.e., O and OH groups), which block active sites for oxygen reduction. As the overpotential is increased, the surface is reduced, the coverage of these species decreases, and the active site density is increased, producing more facile kinetics.

In terms of Pt NP/C catalysts, the problem of oxide adsorption is more significant as a result of their small 2–3 nm size and high defect site density (Koper, 2009). It has been widely shown that spherical nanoparticles display a significant size-dependent increase in surface energy as particle size is decreased below 10 nm (Tang & Cheng, 2015). Higher surface energies contribute to stronger binding energies with adsorbates and thus, larger overpotentials are necessary to reduce the surface coverage of oxide species. Moreover, as particle size is decreased, the relative density of low coordination atoms (LCAs) located at the edges and vertices of the particle's surface increases relative to the smooth defect-free crystalline planes of the facets. LCA and other defect sites strongly adsorb oxide species and further reduce the catalytic activity of the particles at low overpotentials. From these collective effects, the specific activity of commercial Pt NP/C is typically observed to be ~ 0.2 mA cm^{-2}, which is significantly lower than that of bulk platinum and of Pt (111) single crystals.

In addition to challenges with reducing oxygen, a key challenge in the development of catalysts for ORR is the durability of nanostructured Pt (Cao, Wu, & Cao, 2014; Shao, Yin, & Gao, 2007; Shao-Horn et al., 2007; Yu & Ye, 2007). During potential cycling, Pt NP/C tend to undergo particle dissolution and ripening leading to a reduction in active surface area. Additionally, dissolution of surface platinum results in a loss of electrical contact between platinum particles and carbon support, further reducing the surface area of Pt. Under half-cell conditions, accelerated durability tests (ADTs) involving commercial Pt NP/C result in a loss of between 40% and 50% of their initial electrochemically active surface area (ECSA) and catalytic activity.

The technological challenges and catalytic shortfalls associated with Pt NP/C has driven a considerable interest in designing more effective electrocatalysts by tuning the structure and composition on the nanoscale (Antolini & Perez, 2011; Morozan et al., 2011; Nie, Li, & Wei, 2015; Wang, Lei, & Guo, 2016; You, Yang, Ding, & Yang, 2013). One strategy has been to tailor the morphology of the catalysts particles with the overarching goal of achieving predictable control over the exposed surface facets and the defect site density (Wang, Daimon, Onodera, Koda, & Sun, 2008). Among the plethora of structures that have been developed, 1D structures such as nanowires (NWs) and nanotubes (NTs) have garnered significant interest as a promising structural paradigm (Koenigsmann & Wong, 2011; Koenigsmann, Scofield, Liu, & Wong, 2012; Lu, Du, & Steinberger-Wilckens, 2016a; Wang, Lv, et al., 2016). A primary advantage of 1D structures is the inherent structural anisotropy, which results from the selective growth of the structure along a designated crystallographic direction. In the case of platinum, growth along the <111> crystallographic direction enables selective display of the low energy Pt {100} and Pt {111} facets, which are both highly active for ORR. In addition, the structural anisotropy has been shown to produce structures with lower defect site densities relative to 0D counterparts. These structural attributes mutually contribute to a weaker interaction of the surface with oxygen adsorbates and culminate in higher achievable catalytic activity and durability.

Experimental evidence for a structure-dependent enhancement in catalytic activity with 1D Pt nanostructures was first observed in 2007 (Chen, Waje, Li, & Yan, 2007) and subsequently there has been incredible growth in the design of 1D electrocatalysts. Early reports from 2007 to 2010 focused largely on elemental Pt and Pd nanostructures and consistently demonstrated that 1D nanostructures maintain higher catalytic activity and durability relative to commercial 0D nanoparticle analogs (Antolini & Perez, 2011; Koenigsmann & Wong, 2011; Morozan et al., 2011). Since 2010, the emphasis has shifted to increasing the structural complexity of 1D catalysts by judiciously modifying the chemical composition and the structural architecture of the catalysts themselves (Koenigsmann, Scofield, et al., 2012; Lu et al., 2016a; Wang, Lv, et al., 2016). Variation of the composition of the catalyst, not only has the potential for decreasing the precious metal content and overall production cost, but also provides for an unprecedented ability to finely tune the structural and electronic property of the ORR active sites. For example, a wide range of NWs and NTs have been synthesized consisting of Pt and Pd combined with first-row transition metals (i.e., Fe, Co, Ni, and Cu) and other precious metals (Au, Ag, and Ir). In addition, recent reports have also successfully demonstrated the formation of hierarchical structures wherein composition is controlled spatially to form core-shell, de-alloyed, dendritic, and porous nanostructures.

In this chapter, we review the recent experimental and computational progress in the development of 1D electrocatalysts architectures for oxygen reduction. We focus on exploring the fundamental structure–function relationships, including the effect of catalyst dimensions, composition, and surface structure on the corresponding catalytic activity and durability. In addition, we also explore the growing use of computational methods and in situ spectroscopic techniques to elucidate the origin of enhanced performance in 1D structures.

4.1 ONE-DIMENSIONAL Pt-BASED CATALYSTS FOR THE OXYGEN REDUCTION REACTION

4.1.1 One-Dimensional Pt Nanostructured Catalysts for the Oxygen Reduction Reaction

Since the first reports of 1D electrocatalysts in the late 2000s, considerable effort has been devoted to the development and study of 1D elemental Pt electrocatalysts. Rational investigation of 1D Pt electrocatalysts has provided important insights into the size- and structure-dependent catalytic properties of 1D structures without the additional complexities of composition effects in alloy-type structures. Elemental Pt electrocatalysts have also played an important role in elucidating the structural evolution and mechanism of degradation of 1D nanostructures under the operating conditions of fuel cell devices (Lu et al., 2016a; Wang, Lv, et al., 2016). Collectively, these fundamental insights have driven the rational development of more complex alloy-type and hierarchical 1D architectures with lower Pt loadings, and better overall activity and durability.

A key advantage of 1D architectures in terms of ORR catalysis was discovered via an investigation of the size-dependent catalytic activity in free-standing Pt NWs. Computational investigations predicted that decreasing the diameter of precious metal nanowires into the "ultrathin regime" of 1–2 nm would produce a contraction of the surface atoms as a result of high degree of strain within the structure itself (Fiorentini, Methfessel, & Scheffler, 1993; Haftel & Gall, 2006; van Beurden & Kramer, 2004). In terms of catalysis, contraction of the active Pt surface atoms was predicted to contribute to a weakening of the interaction of adsorbates and more facile ORR kinetics (Matanović, Kent, Garzon, & Henson, 2012, 2013).

On the basis of the computational work, Wong and coworkers prepared free-standing, ultrathin Pt NWs (Fig. 4.1B) with an average diameter of 1.3 nm via a solution-based method. The synthesis incorporated an acid-etch treatment to remove defect sites from the surfaces of the wires (Koenigsmann, Scofield, et al., 2012; Koenigsmann, Zhou, Adzic, Sutter, & Wong, 2010). In addition, larger Pt NTs and NWs (Fig. 4.1A) with predictable average diameters of 200 and 45 nm, respectively, were prepared via a template-based method. Detailed investigation of the size-controlled NWs utilizing high-resolution transmission electron microscopy (HRTEM) confirmed that in all cases, the surfaces

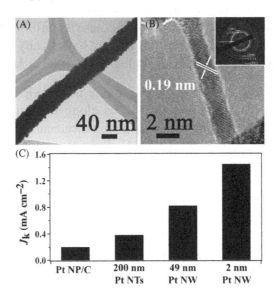

Figure 4.1 Representative TEM images collected from individual isolated Pt NWs possessing diameters of 49 nm (A) and 2 nm (B). A representative selected area electron diffraction pattern collected from an ensemble of individual 2 nm NWs is shown as an inset. The size-dependent trend in 1D Pt nanostructures of specific ORR activity measured at 0.9 V with commercial Pt NP/C serving as a commercial reference system is illustrated in (C). Adapted with permission from Koenigsmann, C., Scofield, M.E., Liu, H., & Wong, S.S. (2012). Designing enhanced one-dimensional electrocatalysts for the oxygen reduction reaction: Probing size- and composition-dependent electrocatalytic behavior in noble metal nanowires. *Journal of Physical Chemistry Letters, 3,* 3385–3398. Copyright 2012 American Chemical Society.

of the NWs were primarily bound by the Pt {100} and Pt {111} facets. The ultrathin Pt NWs were found to have an exceptionally high specific activity of 1.45 mA cm^{-2} at 0.9 V and Koutecky–Levich analysis suggested that the catalysis largely followed the desirable four-electron process. In terms of the catalyst diameter, the specific activity (Fig. 4.1C) was found to increase by a factor of nearly four-fold from 0.38 to 1.45 mA cm^{-2} as the wire diameter was decreased from 200 to 1.3 nm. Analysis of the cyclic voltammetry indicated a positive shift of 10 mV in the position of the oxide reduction peak, suggesting the enhanced activity could be attributed to a progressively weaker interaction with the oxygen adsorbate as a function of decreasing wire diameter. Further evidence for a shift in the d-band center was also observed by Guo and coworkers who detected a shift in the Pt 4f$_{7/2}$ XPS peak consistent with a favorable down-shift in the energy of the d-band center (Wang, Jiang, Wang, & Guo, 2011).

The remarkable size-dependent catalytic activity of Pt NWs highlights a key benefit of 1D systems over 0D since the increase in activity is coupled with an increase in the specific surface area, as particle size is decreased. Since the initial report of Wong and coworkers, other reports have observed enhanced specific activity in ultrathin Pt NWs prepared by solution-based methods and by biologically inspired methods (Ruan et al., 2013; Xiao, Cai, Balogh, Tessema, & Lu, 2012).

Faceted 1D nanostructures also provide an excellent platform for selectively displaying highly active facets along the anisotropic direction of the wire, enabling significant enhancements in performance (Bu, Feng, et al., 2016; Xu et al., 2016). For example, Huang and coworkers examined the facet-dependent ORR activity in well-defined Pt NWs and Pt octahedra bound by Pt {111} facets (Bu, Feng, et al., 2016). The Pt nanostructures were synthesized via a solution-based method wherein platinum precursors were reduced by glucose in the presence of amine-based surfactants. The amine-based surfactants employed in the synthesis strongly adsorbed to the Pt {111} facet, enabling selective display of this facet at the surface of the particles. Interestingly, the degree of structural anisotropy in the resulting particles (i.e., nanowires vs. nanoctahedra) was achieved by varying the relative hydrophobicity of the Pt precursor employed in the synthesis. The specific and mass activities of the as-synthesized Pt NWs were determined to be 2.16 mA cm^{-2} and 0.49 A mg^{-1}, which were significantly higher than the activity of 1.7 mA cm^{-2} and 0.30 A mg^{-1} for the Pt octahedra. Since both structures were bound by Pt {111} facets, the improved performance in the NWs was attributed to the relative defect site density in the structures. In this case, HRTEM analysis of the NWs indicated that they had fewer edge and corner defect sites than the symmetrical octahedra.

Elemental, 1D Pt nanostructures have also served as a platform for elucidating the mechanism of particle degradation and activity loss under operating conditions. Typically, the durability of nanostructured catalysts is examined under half-cell conditions, wherein the catalyst is supported on a planar electrode surface. The potential is cycled repeatedly to simulate the effects of "start-stop" driving on a catalyst within an operational automotive fuel cell. For example, in the aforementioned study by Huang and coworkers, the relatively low-defect site density of the Pt NWs relative to the octahedra contributed to better performance of the faceted Pt NWs in an ADT (Bu, Feng, et al., 2016). In the case of the faceted nanoparticles, the Pt NWs retained nearly 90% of the initial mass activity after 30,000 cycles of the ADT, whereas the octahedra maintained 60% of the initial mass activity. The evolution of the morphology of the octahedra suggested that defect sites at the edges and corners were etched, producing a concave structure. On the other hand, the morphology of the Pt NWs was essentially unaffected during the ADT.

Catalyst degradation was also examined in a series of tubular Pt nanostructures prepared by chemical deposition of Pt onto the surface of fivefold twinned Pd NWs(Wang et al., 2015). In this case, the degree of surface LCA and defect sites in the 1D Pt tubular structures could be tuned via a postsynthetic heat treatment. As-prepared Pt NTs consisted of interconnected Pt NPs with relatively high-defect site densities. Upon a brief heat treatment at 250°C, the particles consolidated into a well-ordered skeletal structure along the edges of the Pd NW template producing a tubular structure with a much lower defect site density. Over the course of an ADT, the defect-rich, as-prepared Pt NTs underwent significant structural reconfiguration and lost nearly 40% of their initial ECSA. On the other hand, the relatively pristine, heat-treated NTs lost only 20% of their initial ECSA and largely retained their morphology. Surprisingly, the structure of the defect-rich Pt NTs evolved into a faceted tubular structure analogous to that of the heat-treated structures. These results suggested the primary mechanism of degradation in the 1D structures was the etching and removal of defect sites.

Moving beyond half-cell testing, there has been a growing interest in examining 1D Pt nanostructures within operating PEMFCs. For example, several reports have demonstrated the successful incorporation of Pt NWs into fuel cell electrodes via the direct growth of Pt NWs on carbon utilizing formic acid as a reducing agent (Du & Pollet, 2012; Li, Higgins, et al., 2015; Meng et al., 2015; Su, Sui, et al., 2014; Su, Yao, et al., 2014). Free-standing Pt NWs synthesized by solution-based methods or by electrospinning have also been successfully incorporated into fuel cell cathodes (Su et al., 2015; Sung, Chang, & Ho, 2014). The transition of catalysts from half-cell testing into operating PEMFCs is encouraging and further highlights the potential of 1D catalysts as plausible alternatives for traditional 0D catalysts. Chapter 6, Proton Exchange Membrane Fuel Cells Electrodes From One-Dimensional Nanostructures, provides a detailed discussion of the recent advances in 1D catalysts incorporated within functional PEMFC electrodes.

4.1.2 Pt-Based Alloy Catalysts for the Oxygen Reduction Reaction

Tailoring the composition of Pt-based catalysts has emerged as a promising approach to improving both the activity and durability of 1D catalysts. A broad range of transition metal elements can be uniformly combined with Pt to form alloys or intermetallic phases. The addition of other elements, especially less expensive first-row transition metals, offsets the high cost of Pt. Beyond simply replacing Pt, combining Pt with other metals provides an unprecedented opportunity to finely tune the electronic and structural properties of the active sites for increased ORR activity.

The activity of 1D Pt nanostructures has been enhanced beyond that of pure Pt by combining Pt with other precious metals such as Pd, Au, and Ag (Chang et al., 2016; Liang et al., 2015; Liu et al., 2016; Lu, Du, & Steinberger-Wilckens, 2016b; Lu, Jiang, & Chen, 2013; Wu et al., 2016; Yeh, Liu, Chen, & Wang, 2013). For example, a recent report by Wang and coworkers has combining electrocatalytic measurements with density functional theory (DFT) to elucidate the structural and electronic effects of precious metals (M = Au, Ag, and Pd) on the catalytic activity and durability of $Pt_{1-x}M_x$ alloy-type nanorods (Liang et al., 2015). In terms of catalytic activity, the measured ORR activities at 0.85 V were found to be 117.2, 102.7, and 103.1 mA cm^{-2} for the PtAu, PtAg, and PtPd alloys, respectively. On the basis of DFT calculations, the trend in activity was attributed to the relative adsorption energy of surface bound −O and −OH intermediates, denoted as $\Delta E_{ads}(O^*)$ and $\Delta E_{ads}(OH^*)$, respectively. Fig. 4.2 highlights the correlation between the activity of the PtM NRs and the calculated values for $\Delta E_{ads}(O^*)$ and $\Delta E_{ads}(OH^*)$. Among the metals in the series, Au is the least oxophilic and contributed to a reduction in the adsorption energy of both −O and −OH surface species. In the case of PtAg, the interaction with surface bound −O decreased but the interaction with surface bound −OH species was not affected. The addition of Pd into the alloy had no significant effect on the adsorption strength of either −O or −OH surface species and thus, maintained the lowest initial specific activity.

Although the PtAu NRs had a high initial activity, the trend in durability after an ADT was found to be PtAg > PtAu > PtPd. Specifically, the PtAg nanorods retained 91% of their initial activity,

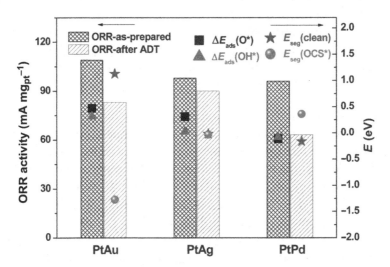

Figure 4.2 A comparison of $\Delta E_{ads}(O^*)$, $\Delta E_{ads}(OH^*)$, $E_{seg}(clean)$, $E_{seg}(OCS^*)$ *and ORR activity before and after 1000 cycles of ADT for PtM alloys.* Adapted from Liang, Y.-T., Lin, S.-P., Liu, C.-W., Chung, S.-R., Chen, T.-Y., ... Wang, K.-W. (2015). The performance and stability of the oxygen reduction reaction on Pt-M (M = Pd, Ag and Au) nanorods: An experimental and computational study. *Chemical Communications, 51,* 6605–6608 with permission from the Royal Society of Chemistry.

while the PtAu and PtPd nanorods retained only $\sim 75\%$ and $\sim 66\%$ of their initial activity, respectively. The significant loss in the activity of the PtAu NWs was a surprising result, since Au is less reactive than Ag and Pd. To elucidate the trend in durability, the surface segregation energy was calculated (Fig. 4.2) for the bare surfaces, E_{seg}(clean), and for the surfaces with adsorbed oxygen containing species, E_{seg}(OCS*). Since Pt atoms are more oxophilic than Au atoms, it is favorable for Pt atoms to be drawn to the surface of the PtAu nanorods when the surface is oxidized, producing a less active Pt-rich surface. On the other hand, DFT calculations predicted that segregation of Pt was much less favorable in the PtAg nanorods, since Ag surface sites adsorb hydroxyl species at operating potentials. Thus these results highlight the complex and sometimes counterintuitive interplay between the composition of a 1D catalysts and its catalytic properties.

Despite the poor durability of the PtAu alloy in earlier studies, Zhong and coworkers have recently demonstrated that the durability of 1D PtAu nanostructures can be enhanced by tuning the surface structure (Chang et al., 2016). Utilizing a hydrothermal method, ultra-thin, helical $Pt_{1-x}Au_x$ alloy-type NWs were prepared, with the gold content ranging from 10% to 84%. The unique helical structure of the NWs selectively displayed the active Pt {111} facet as opposed to the mixture of Pt {111} and Pt {100} facets observed in other reports. The optimized catalyst containing 25% Au maintained a promising ORR mass activity of 0.5 A mg^{-1}. In this case, the polarization curve remained essentially unchanged after 10,000 cycle ADT, indicating that there was no change in the catalytic activity of the helical NWs.

Among the precious metals, the system that has been most studied in 1D architectures is the $Pt_{1-x}Pd_x$ alloy. In one study, the effect of composition on the catalytic activity was examined in a series of $Pt_{1-x}Pd_x$ NWs produced by an ambient, surfactantless, template-based method (Koenigsmann, Sutter, Chiesa, Adzic, & Wong, 2012). The activity of $Pt_{1-x}Pd_x$ NWs increased from 0.64 to 0.79 ± 0.01 mA cm^{-2} as the Pt content was increased from 50% to 80%, respectively. Although the activity did not surpass that of 0.82 mA cm^{-2} for pure Pt NWs, the PtPd NWs maintained an activity that was 80% that of pure Pt NWs despite containing only 50% Pt. The addition of Pd was also found to greatly improve the tolerance of the catalyst to small organic molecules such as methanol. In fact, the PtPd NWs retained

80% of their initial activity upon increasing the concentration of methanol to 4 mM, while the Pt NWs retained only 40% of their initial activity (Koenigsmann & Wong, 2013). Similar results were noted with dendritic PtPd nanostructures prepared by Lee and coworkers, which showed a high tolerance to ethanol (Li, Gao, Li, Chen, & Lee, 2015).

In terms of size-dependent performance, Tang and coworkers prepared PtPd NWs via a surfactant-based solution method with ultrathin diameters of 4.1 nm (Wu et al., 2016). In this case, the ECSA of the NWs was found to be 22.3 $m^2 g^{-1}$, which was larger than that of commercial Pt NP/C or Pd NP/C ($\sim 16\ m^2 g^{-1}$). The activity of the ultrathin wires was also four-fold greater than that of the Pt NP/C when measured at 0.85 V. A noteworthy finding in this report was that the ultrathin PtPd NWs maintained an excellent durability and after an ADT, the ultrathin NWs had essentially no loss in either the measured half-wave potential or the ECSA.

A second route to achieving higher catalytic activity in 1D Pt-based catalysts is the incorporation of first-row transition metals (M = Fe, Co, Ni, Cu, etc.) into binary $Pt_{1-x}M_x$ or ternary formulations (Bu et al., 2015; Cui, Li, Liu, Gao, & Yu, 2012; Guo et al., 2015; Higgins et al., 2014; Liu, Xu, et al., 2015; Wittkopf, Zheng, & Yan, 2014; Yang, Yan, et al., 2016; Yang, Jin, et al., 2016). Although initially considered to a be a disadvantage in terms of catalyst durability, several groups have taken advantage of the instability of first-row transition metal alloys in acidic solution to form highly active 1D catalysts via so-called de-alloying processes. For example, Yu and coworkers systematically investigated the origin of enhanced performance in de-alloyed 1D PtCu nanostructures (Cui et al., 2012). In this case, PtCu NTs were prepared via electrodeposition within an anodic alumina template and the as-prepared wires were subsequently heat treated to vary the relative degree of Pt content at the surface of the NTs. Upon cycling the potential of the various PtCu NTs in acidic electrolytes, the surfaces underwent significant restructuring (cf. Fig. 4.3A and B) due to selective dissolution of surface and subsurface Cu atoms. Restructuring of the catalytic interface was significant in the catalyst samples denoted as "CP3" and "CP5," which were annealed at relatively low temperatures of 300°C and 500°C, respectively. These catalysts were found to have roughened defect-rich surfaces, with activities below 0.4 mA cm^{-2}. The catalysts prepared at higher annealing

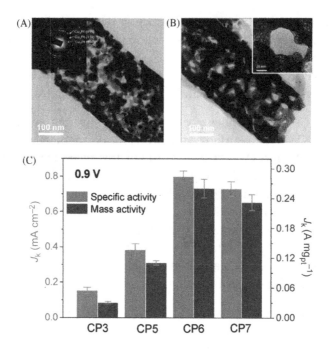

Figure 4.3 (A) TEM images of a PtCu tube after annealing at 600°C. The inset is an electron diffraction pattern of the wire. (B) TEM image of a PtCu restructured tube after 250 cycles between 0.05 and 1.2 V versus reversible hydrogen electrode (RHE) in Ar-saturated 0.1 M HClO₄ solutions. The inset depicts an amplified TEM image showing the roughened surface. (C) Specific and mass activities of PtCu at 0.9 V versus RHE. Specific and mass activities are depicted as kinetic current densities normalized to the ECSA and Pt mass, respectively. Adapted with permission from Cui, C.-H., Li, H.-H., Liu, X.-J., Gao, M.-R., & Yu, S.-H. (2012). Surface composition and lattice ordering-controlled activity and durability of CuPt electrocatalysts for oxygen reduction reaction. ACS Catalysis, 2, 916–924. Copyright 2012 American Chemical Society.

temperatures of 600°C and 700°C denoted as "CP6" and "CP7," respectively, were comparatively stable upon cycling in the electrolyte and evinced activities above 0.7 mA cm^{-2}. The optimum specific activity and mass activity (Fig. 4.3C) of 0.8 mA cm^{-2} and 0.2 A mg^{-1} was achieved with an annealing temperature of 600°C. The increase in activity at higher annealing temperatures was attributed to the heat treatment, which enriched the Pt content at the catalytic interface and increased the order of the PtCu lattice.

Recently, Yan and coworkers employed a similar approach to prepare ordered PtCu alloys at the surface of Cu NWs (Wittkopf et al., 2014). In this case, the optimized catalyst possessed a mass activity and specific activity of 1.06 A mg^{-1} and 2.30 mA cm^{-2}, which was attributed to the optimal degree of compressive strain induced by the

PtCu alloy. In a different report, Huang and coworkers synthesized $Pt_{1-x}Co_x$ NWs that were selectively grown to display higher energy Pt (310) and Pt (110) facets (Bu, Guo, et al., 2016). The specific activity of the wires was found to increase from $\sim 0.5\,mA\,cm^{-2}$ to an outstanding value of $7.12\,mA\,cm^{-2}$ as the cobalt content was increased from 9% to 25%. The optimized Pt_3Co catalyst was also found to have a high platinum mass activity of $3.71\,A\,mg^{-1}$. In this case, the high activity was found to result from highly active hollow sites present on the alloy's high index facets. Similar results were also observed by the same group in $Pt_{1-x}Ni_x$ NWs bound by high index facets (Bu et al., 2015).

4.1.3 One-Dimensional Pt-Based Hierarchical Catalysts for the Oxygen Reduction Reaction

The design of hierarchical structures has enabled control, not only over the elements present and their relative quantities, but also control over the spatial distribution of elements throughout the nanostructure itself. In terms of catalysis, the design of Pt-based hierarchical catalysts has focused on localizing Pt at the surface of the nanostructure leading to core-shell and dendritic architectures. Isolating Pt at the catalytic interface has the potential of greatly reducing the necessary Pt content, since the catalytically inactive core of the material can be replaced with less expensive and more abundant elements. In addition, the inherent structural and electronic interactions between the surface Pt atoms and the atoms at the core of the material can be tuned to increase the catalytic activity and durability of the Pt itself.

Core-shell structures are a promising route to achieve highly active catalysts with low platinum loadings, since the Pt is localized at the NW or NT as a thin shell. For example, DFT calculations performed by Norskøv and coworkers examined the effect of depositing a monolayer of Pt atoms atop various transition metal substrates (Nørskov et al., 2004). The results predicted a "volcano-type" behavior with the peak of activity achieved with substrates, such as Pd, that have slightly smaller lattice constants than Pt. The strain induced by the Pd core on the Pt produces a contraction of the Pt atoms, contributing to a down-shift in the d-band center of the Pt surface sites, which leads to weaker binding energies of Pt active sites with adsorbed oxygen species.

In light of these predictions, several groups have developed methodologies to prepare thin Pt shells onto Pd NW cores. For example, a platinum monolayer was deposited onto Pd NWs with diameters ranging from 270 to 2 nm via a two-step Cu underpotential deposition (UPD) process (Koenigsmann, Santulli, Gong, et al., 2011; Koenigsmann, Santulli, Sutter, & Wong, 2011). The activity of the resulting core-shell NWs revealed a significant size-dependent enhancement in the activity of the ultrathin NWs (0.74 mA cm^{-2} and 1.74 A mg$_{Pt}^{-1}$), when compared with the larger 45 nm (0.55 mA cm^{-2} and 1.28 A mg$_{Pt}^{-1}$) or 270 nm (0.40 mA cm^{-2} and 1.01 A mg$_{Pt}^{-1}$). Recently, Yu and coworkers have demonstrated the successful synthesis of thin Pt shells with thicknesses of 0.5−0.8 nm via a solution-based process (Li, Ma, et al., 2015). Although the Pt mass activity of the wires (0.8 A mg^{-1}) was lower than the activity of the wires prepared by Cu UPD, solution-based methods for forming thin Pt shells are desirable in terms of scalability.

In an effort to tune the electronic and structural interactions between the core NW and the Pt shell, Wong and coworkers prepared a series of hierarchical, 1D catalysts consisting of a single atomic layer of Pt deposited onto ultrathin, Pd$_{1-x}$M$_x$ (M = Au and Ni) alloy NW cores (Koenigsmann, Santulli, Gong, et al., 2011; Koenigsmann, Santulli, Sutter, et al., 2011; Koenigsmann, Sutter, Chiesa, et al., 2012; Koenigsmann, Sutter, Adzic, & Wong, 2012; Liu, An, et al., 2015; Liu, Koenigsmann, Adzic, & Wong, 2014). The ultrathin Pd$_{1-x}$M$_x$ NWs ($d = 2$ nm) were synthesized in the presence of octadecylamine to selectively display the active Pd {100} facet. After removing the surfactant, a conformal atomic layer of Pt was deposited on the surface of these NWs via Cu UPD. Interestingly, in both cases, the activity was found to follow a volcano-type behavior with a peak activity achieved with the Pd$_9$M alloys. In the case of the Pd$_{1-x}$Au$_x$ alloys, the specific activity and platinum mass activity were found to increase from 0.61 to 0.98 mA cm^{-2} and 1.21 to 2.54 A mg$_{Pt}^{-1}$, respectively, as the gold content decreased from 30% to 10%. Among the two metals examined, the gold additive produced the highest ORR activity (0.98 mA cm^{-2}, 2.54 A mg^{-1}), which was approximately 1.5-fold higher than that of the Ni-based NWs (0.62 mA cm^{-2}, 1.44 A mg^{-1}). In terms of durability, both catalysts were found to retain their initial activity over the course of 10,000 cycles of an ADT.

The Pt monolayer supported on the alloy NW cores significantly outperformed the analogous catalyst consisting of a Pt monolayer supported on a pure Pd core. This highlights the intrinsic benefits of coupling the surface Pt atoms with a core NW optimized in terms of its composition. To investigate the origin of enhanced activity in these unique hierarchical core-shell structures, the structure of the catalysts was examined by in situ extended X-ray absorption fine structure (EXAFS) measurements (Liu, An, et al., 2015). The EXAFS results revealed that, during the Cu UPD process, Au atoms are segregated at the surface of the wire, leading to a PtAu monolayer and a predominantly pure Pd core. DFT calculations were performed on several plausible structures and the surface segregated structure observed by EXAFS was the only structure that fully reproduced the volcano-type behavior in activity. Based on the calculations, the segregation of Au toward the surface contributes to a higher stability of the wire and also weakens the binding energy of oxygen adsorbates, thereby explaining the enhanced performance.

Sun and coworkers have recently demonstrated that the results with binary alloy cores can be extended to ternary alloy cores with Pt-based shells (Guo, Zhang, Su, & Sun, 2013). In this case, ternary FePtM (M = Pd or Au) were synthesized via a solution-based method and were coated with a thin FePt shell via a seed-mediated thermal decomposition process. The specific activity and mass activity (Fig. 4.4A) of the resulting FePt-coated FePtPd NWs reached $3.47 \, \text{mA cm}^{-2}$ and $1.68 \, \text{A mg}_{\text{Pt}}^{-1}$ for the NW coated with a 0.8 nm thick shell. At this same potential, a commercial Pt NP/C catalyst maintained a specific activity of only $0.24 \, \text{mA cm}^{-2}$. Interestingly, the FePtPd core measurably outperformed the FePtAu ($1.59 \, \text{mA cm}^{-2}$) by nearly two-fold. Although the origin of enhancement remains unclear, the enhanced activity is attributed to the different electronic effects associated with Pd and Au in the hierarchical structure. It is also plausible that dissolution of iron from the structure and subsequent segregation of the atoms may contribute to the enhanced performance in the case of the FePtPd alloy core. In terms of durability, the polarization curve (Fig. 4.4B) and morphology (Fig. 4.4C. and 4.4D) of the optimized FePtPd/FePt NWs remained essentially identical after a 5000-cycle ADT.

Moving toward the use of more abundant and less expensive metals, Yan and coworkers have developed a class of highly active 1D

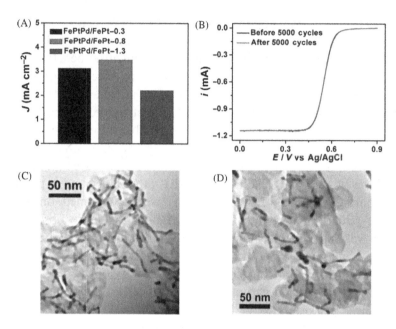

Figure 4.4 (A) Summary of the specific activities for ORR at 0.5 V versus Ag/AgCl (4 M KCl) for FePtPd/FePt core-shell NWs with shell thicknesses of 0.3, 0.8, and 1.3 nm. (B) Polarization curves of the FePtPd/FePt NWs with 0.8 nm thick shells before and after 5000 potential cycles between 0.4 and 0.8 V. (C, D) TEM images of the FePtPd/FePt NWs with 0.8 nm thick shells before (C) and after (D) the stability test. Adapted with permission from Guo, S., Zhang, S., Su, D., & Sun, S. (2013). Seed-mediated synthesis of core/shell FePtM/FePt (M = Pd, Au) nanowires and their electrocatalysis for oxygen reduction reaction. *Journal of the American Chemical Society, 135,* 13879–13884. Copyright 2013 American Chemical Society.

catalysts consisting of Pt shells deposited on Cu, Co, and Ni NW cores (Alia et al., 2013; Alia, Larsen, et al., 2014; Alia, Pylypenko, et al., 2014; Alia, Yan, & Pivovar, 2014). Uniform Pt shells are deposited onto the surface of the NWs via spontaneous galvanic displacement, wherein a Pt precursor is reduced by the sacrificial oxidation of the Cu, Co, or Ni atoms at the surface of the NW. Exposure of the catalysts to potential cycling in the perchloric acid electrolyte resulted in significant structural change of the core-shell structure in the case of the Cu and Co core-shell wires, leading to a porous structure. On the other hand, the core-shell wires with Ni NW cores largely retained the core-shell morphology even after 30,000 cycles of an ADT. In terms of catalytic activity, the mass activities for the Ni, Co, and Cu NWs with optimized compositions were 0.92, 0.79, and 0.45 A mg^{-1}, respectively. The contrasting trends in mass activity and specific activity suggested that the Pt and Ni, Co, or Cu content were largely segregated with minimal alloying between the two components.

In addition to core-shell nanostructures, a second route to achieving enhanced activity has focused on forming complex three-dimensional (3D) architectures derived from one-dimensional (1D) motifs. In practice, ORR catalysts have been formed by depositing Pt or Pt-alloy clusters or dendrites onto 1D substrates consisting of metals, metal oxides, or semiconductors. In one report, Yin and coworkers decorated Pd nanorods with $1-2$ nm Pt clusters via a modified Cu UPD process leading to a catalyst with a low overall Pt content of 1.5% (Yang, Cao, et al., 2015). In this case, the UPD process was performed entirely in solution by utilizing a weak reducing agent to selectively reduce copper at the surface of the Pd substrate. Subsequent galvanic displacement led to the formation of small Pt clusters dispersed across the surface of the rods. The measured specific activity (0.18 mA cm^{-2}) was comparable with that of Pt NP/C, however the very low Pt loading engendered by the hierarchical structure led to a Pt mass activity of 7 A mg$_{Pt}^{-1}$. In another report, a similar architecture was prepared by depositing Pt clusters onto PtTe alloy NWs (Fig. 4.5A) via a solution-based process (Li, Xie, et al., 2016). The specific and mass activity (Fig. 4.5B) of the

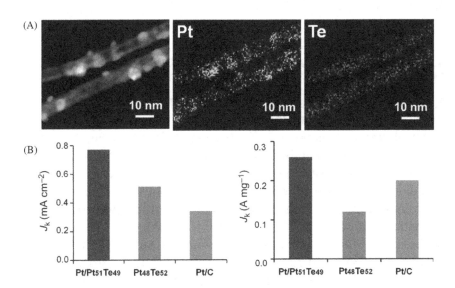

Figure 4.5 (A) High-angle annular dark field (HAADF) images (left) and energy dispersive X-ray spectroscopy images (right) indicating the distribution of the Pt and Te elements throughout the Pt/Pt$_{51}$Te$_{49}$ NWs. (B) Specific (left) and mass (right) activities of the hierarchical Pt/Pt$_{51}$Te$_{49}$ NWs, Pt$_{51}$Te$_{49}$ NWs, and commercial Pt NP/C. Adapted with permission from Li, H.-H., Xie, M.-L., Cui, C.-H., He, D., Gong, M., ... Yu, S.-H. (2016b). Surface charge polarization at the interface: Enhancing the oxygen reduction via precise synthesis of heterogeneous ultrathin Pt/PtTe nanowire. Chemistry of Materials, 28, 8890−8898. Copyright 2016 American Chemical Society.

hierarchical structure was determined to be $\sim 0.8\,\mathrm{mA\,cm}^{-2}$ and $\sim 0.25\,\mathrm{A\,mg}^{-1}$, which was approximately two-fold higher than that of analogous $Pt_{1-x}Te_x$ alloy NWs. DFT calculations revealed that the unique Pt—Te interface in the hierarchical wires led to a much lower activation energy barrier for the desorption of oxygen from the Pt active sites than in the alloys.

Hierarchical structures have also been prepared utilizing carbon nanotubes (CNTs) as 1D supports. Raj and coworkers deposited Pd clusters onto multiwalled carbon nanotubes (MWCNTs), which served as seeds for the growth of Pt dendrites (Ghosh, Mondal, & Retna Raj, 2014). MWCNTs are an interesting alternative to traditional mesoporous, particulate carbon supports since they are more robust and promote lower charge transport resistance due to guided electron transport (Peng & Wong, 2009; Peng, Chen, Misewich, & Wong, 2009). In this case, the specific activity increased from 0.150 to $0.342\,\mathrm{mA\,cm}^{-2}$, respectively, as the composition of Pt increased from 21% to 64%. In addition, the polarization curve of the optimized hierarchical catalysts remained essentially the same over the course of a brief 1000 cycle ADT.

4.2 ONE-DIMENSIONAL NON-Pt-BASED CATALYSTS FOR THE OXYGEN REDUCTION REACTION

4.2.1 One-Dimensional Non-Pt, Precious Metal Catalysts for the Oxygen Reduction Reaction

Although tremendous progress has been made in improving the activity of Pt, the development of Pt-free catalysts is an attractive route to greatly reduce the cost of precious metal catalysts. In acidic conditions, palladium has garnered the most interest as a potential replacement for platinum owing to its similar structural and electronic properties and its dramatically lower cost. For example, Pt and Pd share the same face-centered cubic crystal structure and maintain nearly identical lattice parameters of 3.92 and 3.89 Å, respectively. Despite the structural similarities, nanoparticulate Pd catalysts achieve only 10%—50% of the activity of analogous nanoparticulate Pt catalysts. Both experimental and computational evidence suggest that this is a result of the higher oxophilicity of Pd relative to Pt, which necessarily leads to higher ORR overpotentials (Antolini, 2009). Thus the development of Pd-based catalysts has focused upon tuning the electronic properties of Pd with the overarching goal of reducing the binding energy of oxygen at its surface.

Since the first report of a structure-dependent enhancement in ORR activity in Pd nanorods by Abruña and coworkers (Xiao, Zhuang, Liu, Lu, & Abruña, 2008), there has been an emerging interest in developing Pd-based catalysts with 1D morphologies. Like their Pt-based counterparts, a significant size-dependent enhancement in ORR activity was noted in segmented, ultrathin Pd NWs relative to the activity of larger single-crystalline 45 and 270 nm Pd NWs (Koenigsmann, Santulli, Gong, et al., 2011; Koenigsmann, Santulli, Sutter, et al., 2011). In this case, the activity of the ultrathin Pd NWs was determined to be $3.62 \, \text{mA cm}^{-2}$ at 0.85 V, which was nearly two-fold higher than that of the 45 and 270 nm NWs. As was the case with ultrathin Pt NWs, the enhanced performance in the ultrathin Pd NWs was believed to arise from a contraction of the wire's surface, induced by the anisotropic structure and the high degree of strain. Such a contraction should lead to weaker interactions with adsorbates such as oxygen. To validate this hypothesis, CO stripping voltammograms were obtained and, as expected, the CO stripping peak shifted negatively from 0.925 to 0.906 V as the size was decreased from 270 to 2 nm. The negative shift in the CO stripping potential is consistent with a decrease in the binding energy of the CO adsorbate leading to the observed improvement in the oxidation kinetics.

Moving beyond elemental Pd, the emphasis has shifted to increasing the activity of 1D Pd catalysts beyond that of Pt NP/C by modifying the chemical composition. For example, Wong and coworkers have prepared ultrathin, bimetallic $Pd_{1-x}M_x$ (M = Ni and Au) NWs via a solution-based method (Koenigsmann, Santulli, Sutter, et al., 2011, Koenigsmann, Santulli, Scofield, et al., 2012, Koenigsmann, Sutter, Adzic, et al., 2012, Koenigsmann, Sutter, Chiesa, et al., 2012c, Liu et al., 2014). The synthesis of ultrathin alloy NWs was achieved by employing amine-based surfactants, which strongly adsorbed to the Pd {100} facet leading to anisotropic growth along the Pd <111> crystallographic direction. The growth direction and uniformity of the NWs was confirmed by transmission electron microscopy (TEM) analysis (Fig. 4.6A and B). The strongly adsorbed surfactants were removed by selectively adsorbing CO to the surface of the NWs to

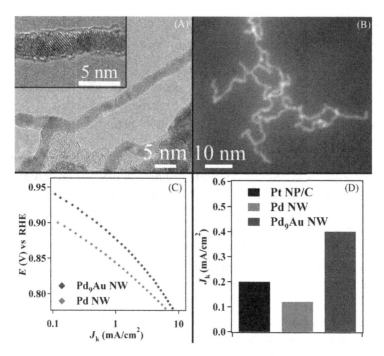

Figure 4.6 High-resolution TEM images (A) of an individual Pd_9Au NW/C and an HAADF image (B) from a representative collection of unsupported NWs. Also shown are the potential versus specific activity (E vs. J_k) plot (C) and the measured area-normalized ORR kinetic current densities (D), which highlight the enhanced performance of the alloyed Pd_9Au NWs by comparison with Pd NWs and Pt NP/C. Reproduced with permission from Koenigsmann, C., Scofield, M.E., Liu, H., & Wong, S.S. (2012). Designing enhanced one-dimensional electrocatalysts for the oxygen reduction reaction: Probing size- and composition-dependent electrocatalytic behavior in noble metal nanowires. *Journal of Physical Chemistry Letters, 3*, 3385–3398. Copyright 2012 American Chemical Society.

displace the surfactant followed by CO stripping to activate the surface (Adzic, Gong, Cai, Wong, & Koenigsmann, 2013).

The performance of the Pd_9Au NWs is summarized in comparison with Pd NWs and Pt NP/C in Fig. 4.5C and D. Notably, the activated Pd_9Au NW evinced a specific activity of 0.4 mA cm^{-2}, which was nearly twofold higher than that of Pt NP/C. Considering the possibility of enhanced performance from bimetallic Pd−Au pair sites at the surface of the wire, the surface of a Pd NW was decorated with Au atoms via galvanic displacement and Cu UPD (Koenigsmann, Sutter, Chiesa, et al., 2012c). In both cases, the activity of the Au modified surface was essentially identical to the pure Pd NW catalysts. Thus the enhanced activity is attributed to the gold atoms in the alloy-type structure, which decrease the oxophilicity of the Pd active sites via a combination of structural and electronic effects. Tooze and coworkers have also achieved promising results with ultrathin PdCu NWs that maintained ORR onset potentials similar to that of Pt (Yu, Zhou, Bellabarba, & Tooze, 2014).

For the first time, Sampath and coworkers have investigated the morphology-dependent ORR activity in Ir electrocatalysts (Chakrapani & Sampath, 2014). Well-defined Ir NWs and NPs with diameters of 2.5 and 3.0 nm, respectively, were prepared via the reduction of $IrCl_3$ in the presence of polyvinylpyrrolidone. The addition of a weak reducing agent (ascorbic acid) to the reaction mixture led to the selective formation of Ir NWs, whereas Ir NPs were formed in the presence of surfactant only. Tafel plots revealed that the slopes in the low current density (-67 mV) and high current density regions (-159 mV) were similar to that of Pt and Pd, suggesting that the mechanism of ORR on Ir is similar to that of Pt and Pd. In addition, Koutecky–Levich analysis of polarization curves collected in 0.5 M H_2SO_4 yielded calculated values for n of ~ 3.6, suggesting that the reaction pathway largely follows a four-electron process at overpotentials between 0.7 and 0.9 V. Although the mechanism was similar for the Ir NPs and NWs, the polarization curve of the NWs was shifted positively by 75 mV relative to the NPs, highlighting a significant enhancement in overall activity. Thus these results suggest that the generally beneficial structural properties of 1D architectures is extended to ORR active metals other than Pt and Pd in acidic media.

Iridium has also been successfully combined with Pd to produce $Pd_{1-x}Ir_x$ NWs with activities that are analogous to commercial Pt NP/C. Yin and coworkers employed a hydrothermal method to coreduce Pd and Ir precursors yielding Pd_2Ir NWs with diameters of 1.5–2.0 nm (Yang, Ma, et al., 2015). Polarization curves collected from electrodes loaded with a fixed quantity of the Pd_2Ir NWs and Pt NP/C indicated that the onset of ORR was slightly delayed in the Pd_2Ir NWs relative to Pt NP/C. However, the half-wave potentials were essentially identical. The mass activity of the as-prepared Pd_2Ir NWs was determined to be 0.12 A mg^{-1} and was essentially identical to that of the Pt NP/C. Despite their similar initial activity, the Pd_2Ir NWs outperformed Pt NP/C over the course of a 10,000 cycle ADT and retained $\sim 75\%$ of their initial activity.

4.2.2 One-Dimensional Nonprecious Metal Catalysts

Nonprecious metal catalysts (NPMCs) are an emerging area of interest and offer the advantage of dramatically lower costs relative to precious metal catalysts. However, the activity and durability of state-of-the-art NPMCs is still far below that of commercial Pt NP/C in acidic media and thus, they are far from commercial viability. Nonetheless, research and development of NPMCs has produced dramatic improvements in catalytic activity by relying upon proven approaches to understanding and improving performance drawn from the development of precious metal catalysts. The potential exists that continued development will lead to NPMC catalysts with performance comparable to or better than Pt NP/C. Herein, we summarize some of the emerging materials that have been investigated with 1D morphologies.

CNTs have garnered a considerable interest as metal-free ORR catalysts since they maintain a measureable ORR activity without the need for metal cocatalysts. Recently, there is a growing interest in tuning the composition of CNTs to achieve higher activity and selectivity for ORR, especially under alkaline conditions (Zhao, Masa, Schuhmann, & Xia, 2013). For example, Giambastiani and coworkers modified the surface of MWCNTs with nitrogen atoms by covalently attaching pyridine functional groups (Tuci et al., 2013). Among the various pyridine groups employed, the MWCNTs functionalized with isopentylnitrile resulted in a catalyst with an activity that nearly matched that of Pt NP/C under alkaline conditions. The enhanced activity is attributed to the polarized N$-$C bond, which results in a weakening of the dioxygen bond upon adsorption and leads to faster overall ORR kinetics.

The activity of CNTs can also be increased by incorporating metal cocatalysts in addition to N-containing surface species. Nakashima and coworkers employed polybenzimidazoles (PBI) to exfoliate MWCNTs and form PBI-coated MWCNT core-shell composites (Fujigaya, Uchinoumi, Kaneko, & Nakashima, 2011). The core-shell structures were then subsequently treated with a solution of Co-ions which coordinated with the N-atoms of the PBI group. In a final step, the Co-modified core-shell structures were heat treated to thermally decompose to the surface layer forming a graphitized structure containing Co and N sites. The activity of the core-shell structure containing Co was found to be significantly higher than that of an analogous structure that underwent an acid treatment to remove the Co atoms. This result suggested that the Co atoms, not only catalyze the graphitization of the PBI during the thermal treatment, but also contribute to the activity of the functioning catalyst.

Transition metal chalcogenide catalysts are also a promising paradigm in NPMCs to achieve higher activity. In one recent example, Cu_2Se NWs catalysts were synthesized by reacting Cu NW templates with a selenium precursor in the presence of surfactants (Liu et al., 2013). The as-prepared NWs could be crystallized in either the tetragonal or cubic Cu_2Se phase by varying the reaction temperature. The different spatial arrangement of the Cu and Se atoms offers a unique opportunity to tune the geometry of the surface sites and the electronic properties of the materials. It was observed that the Cu_2Se NWs synthesized with the tetragonal phase maintained a higher ORR activity and the mechanism was found to largely follow a four-electron process, while the mechanism of the NWs crystallized in the cubic phase followed a combined two- and four-electron process.

Transition metal oxides have also become a focus of interest in the search for NPMCs especially as ORR catalysts in alkaline conditions. Spinel 1D NWs, such as Marokite $CaMn_2O_4$ and $NiCo_2O_4$ are examples of such materials with activities that are comparable to that of a Pt/C catalyst (Du et al., 2012; Jin, Lu, Cao, Yang, & Yang, 2013). There is also a growing interest in manganese oxide owing to its high activity. For example, MnO_2 NWs doped with Ni and Co were prepared through hydrothermal reactions and then blended with graphene-like carbon to form ORR catalysts (Lambert et al., 2012). Manganese oxide has also been deposited onto CNTs via electrodeposition resulting in hybrid structures that benefit from the high activity of metal oxides and the high conductivity of CNTs (Yang, Zhou, Nie, Yao, & Huang, 2011).

4.3 SUMMARY AND OUTLOOK

Since the earliest reports in 2007, the development of 1D, ORR electrocatalysts have undergone profound growth both in terms of the breadth of structures developed, and in terms of the achievable activity and durability. These advances have drawn upon the expanding landscape of structure and composition, which affords an unprecedented opportunity to control the catalytic properties of the material by tuning the electronic and structural properties of the active sites. Although surveying the range of possible compositions has identified many promising catalyst systems, the recent and growing use of in situ spectroscopy and computational methods along with an emergent ability to precisely tune the 3D structure and composition of as-synthesized catalysts have allowed for an unprecedented ability to examine the fundamental correlations between structure and function in these catalysts. Collectively, these advances have enabled a transition from 1D Pt-based systems to catalysts composed of transition metals that are considerably less expensive and more abundant than Pt, yet maintain similar or even better activities than Pt. Although the majority of effort has been dedicated to systems containing other less expensive precious metals like Pd, there is an emerging body of literature that have identified 1D electrocatalysts that do not require any precious metal content, with promising activity and durability. It is likely that the structure—function correlations identified in precious metals along with the ever increasing emphasis on combinatorial approaches to catalyst design will allow for a rapid growth and development of these systems.

In moving toward the future, a key stepping stone for the practical development of 1D ORR electrocatalysts will be to transition from half-cell testing to the examination of promising catalysts in functioning PEMFC devices. The transition toward investigating performance within operating devices will provide important and necessary insights into the performance of 1D catalyst systems within the complex architecture of the device itself, as well as provide much needed insights into the challenges of building electrodes with 1D catalysts. Moreover, further investigation of the performance of 1D systems within fuel cells will allow for a broader understanding of the commercial potential for these systems. Collectively, there is a bright horizon for 1D ORR catalysts and through continued effort it is likely that practical, cost-effective catalysts can be achieved.

ACKNOWLEDGMENTS

Christopher Koenigsmann and Ian Colliard would like to thank Alexander C. Santulli and Erin L. Culbert for their assistance with revising the manuscript and helpful discussions.

REFERENCES

Adzic, R.R., Gong, K., Cai, Y., Wong, S.S., & Koenigsmann, C. (2013). *Method for removing strongly adsorbed surfactants and capping agents from metal to facilitate their catalytic applications.* United States Patent. November 8, 2016.

Adzic, R. R., & Wang, J. X. (1998). Configuration and site of O_2 adsorption on the Pt(111) electrode surface. *Journal of Physical Chemistry B, 102,* 8988–8993.

Adzic, R. R., & Wang, J. X. (2000). Structure of active phases during the course of electrocatalytic reactions. *Journal of Physical Chemistry B, 104,* 869–872.

Alia, S. M., Jensen, K., Contreras, C., Garzon, F., Pivovar, B., & Yan, Y. (2013). Platinum coated copper nanowires and platinum nanotubes as oxygen reduction electrocatalysts. *ACS Catalysis, 3,* 358–362.

Alia, S. M., Larsen, B. A., Pylypenko, S., Cullen, D. A., Diercks, D. R., Neyerlin, K. C., ... Pivovar, B. S. (2014). Platinum-coated nickel nanowires as oxygen-reducing electrocatalysts. *ACS Catalysis, 4,* 1114–1119.

Alia, S. M., Pylypenko, S., Neyerlin, K. C., Cullen, D. A., Kocha, S. S., & Pivovar, B. S. (2014). Platinum-coated cobalt nanowires as oxygen reduction reaction electrocatalysts. *ACS Catalysis, 4,* 2680–2686.

Alia, S. M., Yan, Y. S., & Pivovar, B. S. (2014). Galvanic displacement as a route to highly active and durable extended surface electrocatalysts. *Catalysis Science & Technology, 4,* 3589–3600.

Antolini, E. (2009). Palladium in fuel cell catalysis. *Energy & Environmental Science, 2,* 915–931.

Antolini, E., & Perez, J. (2011). The Renaissance of unsupported nanostructured catalysts for low-temperature fuel cells: From the size to the shape of metal nanostructures. *Journal of Materials Science, 46,* 1–23.

Bu, L., Ding, J., Guo, S., Zhang, X., Su, D., Zhu, X., ... Huang, X. (2015). A general method for multimetallic platinum alloy nanowires as highly active and stable oxygen reduction catalysts. *Advanced Materials, 27,* 7204–7212.

Bu, L., Feng, Y., Yao, J., Guo, S., Guo, J., & Huang, X. (2016). Facet and dimensionality control of Pt nanostructures for efficient oxygen reduction and methanol oxidation electrocatalysts. *Nano Research, 9,* 2811–2821.

Bu, L., Guo, S., Zhang, X., Shen, X., Su, D., Lu, G., ... Huang, X. (2016). Surface engineering of hierarchical platinum-cobalt nanowires for efficient electrocatalysis. *Nature Communications, 7,* 11850.

Cao, M., Wu, D., & Cao, R. (2014). Recent advances in the stabilization of platinum electrocatalysts for fuel-cell reactions. *ChemCatChem, 6,* 26–45.

Chakrapani, K., & Sampath, S. (2014). The morphology dependent electrocatalytic activity of Ir nanostructures towards oxygen reduction. *Physical Chemsitry Chemical Physics, 16,* 16815–16823.

Chang, F., Shan, S., Petkov, V., Skeete, Z., Lu, A., Ravid, J., ... Zhong, C.-J. (2016). Composition tunability and (111)-dominant facets of ultrathin platinum–gold alloy nanowires toward enhanced electrocatalysis. *Journal of the American Chemical Society, 138,* 12166–12175.

Chen, Z., Waje, M., Li, W., & Yan, Y. (2007). Supportless Pt and PtPd nanotubes as electrocatalysts for oxygen-reduction reactions. *Angewandte Chemie International Edition, 46,* 4060–4063.

Chuan-Jian, Z., Jin, L., Bin, F., Bridgid, N. W., Peter, N. N., Rameshwori, L., & Jun, Y. (2010). Nanostructured catalysts in fuel cells. *Nanotechnology, 21,* 062001.

Cui, C.-H., Li, H.-H., Liu, X.-J., Gao, M.-R., & Yu, S.-H. (2012). Surface composition and lattice ordering-controlled activity and durability of CuPt electrocatalysts for oxygen reduction reaction. *ACS Catalysis, 2*, 916–924.

Du, J., Pan, Y., Zhang, T., Han, X., Cheng, F., & Chen, J. (2012). Facile solvothermal synthesis of $CaMn_2O_4$ nanorods for electrochemical oxygen reduction. *Journal of Materials Chemistry, 22*, 15812–15818.

Du, S., & Pollet, B. G. (2012). Catalyst loading for Pt-nanowire thin film electrodes in PEFCs. *International Journal of Hydrogen Energy, 37*, 17892–17898.

Fiorentini, V., Methfessel, M., & Scheffler, M. (1993). Reconstruction mechanism of FCC transition metal (001) surfaces. *Physical Review Letters, 71*, 1051–1054.

Fujigaya, T., Uchinoumi, T., Kaneko, K., & Nakashima, N. (2011). Design and synthesis of nitrogen-containing calcined polymer/carbon nanotube hybrids that act as a platinum-free oxygen reduction fuel cell catalyst. *Chemical Communications, 47*, 6843–6845.

Gewirth, A. A., & Thorum, M. S. (2010). Electroreduction of dioxygen for fuel-cell applications: Materials and challenges. *Inorganic Chemistry, 49*, 3557–3566.

Ghosh, S., Mondal, S., & Retna Raj, C. (2014). Carbon nanotube-supported dendritic Pt-on-Pd nanostructures: Growth mechanism and electrocatalytic activity towards oxygen reduction reaction. *Journal of Materials Chemistry A, 2*, 2233–2239.

Greeley, J., & Nørskov, J. K. (2009). Combinatorial density functional theory-based screening of surface alloys for the oxygen reduction reaction. *Journal of Physical Chemistry C, 113*, 4932–4939.

Guo, H., Liu, X., Bai, C., Chen, Y., Wang, L., Zheng, M., ... Peng, D.-L. (2015). Effect of component distribution and nanoporosity in CuPt nanotubes on electrocatalysis of the oxygen reduction reaction. *ChemSusChem, 8*, 486–494.

Guo, S., Zhang, S., Su, D., & Sun, S. (2013). Seed-mediated synthesis of core/shell FePtM/FePt (M = Pd, Au) nanowires and their electrocatalysis for oxygen reduction reaction. *Journal of the American Chemical Society, 135*, 13879–13884.

Haftel, M. I., & Gall, K. (2006). Density functional theory investigation of surface-stress-induced phase transformations in FCC metal nanowires. *Physical Review B: Condensed Matter, 74*, 035420-035412.

He, C., Desai, S., Brown, G., & Bollepalli, S. (2005). PEM fuel cell catalysts: Cost, performance, and durability. *Electrochemical Society Interface, 14*, 41–44.

Higgins, D. C., Wang, R., Hoque, M. A., Zamani, P., Abureden, S., & Chen, Z. (2014). Morphology and composition controlled platinum–cobalt alloy nanowires prepared by electrospinning as oxygen reduction catalyst. *Nano Energy, 10*, 135–143.

Jin, C., Lu, F., Cao, X., Yang, Z., & Yang, R. (2013). Facile synthesis and excellent electrochemical properties of $NiCo_2O_4$ spinel nanowire arrays as a bifunctional catalyst for the oxygen reduction and evolution reaction. *Journal of Materials Chemistry A, 1*, 12170–12177.

Koenigsmann, C., Santulli, A. C., Gong, K., Vukmirovic, M. B., Zhou, W.-p, Sutter, E., ... Adzic, R. R. (2011). Enhanced electrocatalytic performance of processed, ultrathin, supported Pd–Pt core–shell nanowire catalysts for the oxygen reduction reaction. *Journal of the American Chemical Society, 133*, 9783–9795.

Koenigsmann, C., Santulli, A. C., Sutter, E., & Wong, S. S. (2011). Ambient, surfactantless synthesis, growth mechanism, and size-dependent electrocatalytic behavior of high-quality, single crystalline palladium nanowires. *ACS Nano, 5*, 7471–7487.

Koenigsmann, C., Scofield, M. E., Liu, H., & Wong, S. S. (2012). Designing enhanced one-dimensional electrocatalysts for the oxygen reduction reaction: Probing size- and composition-dependent electrocatalytic behavior in noble metal nanowires. *Journal of Physical Chemistry Letters, 3*, 3385–3398.

Koenigsmann, C., Sutter, E., Adzic, R. R., & Wong, S. S. (2012). Size- and composition-dependent enhancement of electrocatalytic oxygen reduction performance in ultrathin palladium-gold (Pd$_{1-x}$Au$_x$) nanowires. *Journal of Physical Chemistry C, 116*, 15297–15306.

Koenigsmann, C., Sutter, E., Chiesa, T. A., Adzic, R. R., & Wong, S. S. (2012). Highly enhanced electrocatalytic oxygen reduction performance observed in bimetallic palladium-based nanowires prepared under ambient, surfactantless conditions. *Nano Letters, 12*, 2013–2020.

Koenigsmann, C., & Wong, S. S. (2011). One-dimensional noble metal electrocatalysts: A promising structural paradigm for direct methanol fuel cells. *Energy & Environmental Science, 4*, 1161–1176.

Koenigsmann, C., & Wong, S. S. (2013). Tailoring chemical composition to achieve enhanced methanol oxidation reaction and methanol-tolerant oxygen reduction reaction performance in palladium-based nanowire systems. *ACS Catalysis, 3*, 2031–2040.

Koenigsmann, C., Zhou, W.-p, Adzic, R. R., Sutter, E., & Wong, S. S. (2010). Size-dependent enhancement of electrocatalytic performance in relatively defect-free, processed ultrathin platinum nanowires. *Nano Letters, 10*, 2806–2811.

Koper, M. T. M. (Ed.), (2009). *Fuel cell catalysts* Hoboken, NJ: Wiley Interscience.

Lambert, T. N., Davis, D. J., Lu, W., Limmer, S. J., Kotula, P. G., Thuli, A., ... Tour, J. M. (2012). Graphene-Ni-α-MnO$_2$ and -Cu-α-MnO$_2$ nanowire blends as highly active non-precious metal catalysts for the oxygen reduction reaction. *Chemical Communications, 48*, 7931–7933.

Li, B., Higgins, D. C., Xiao, Q., Yang, D., Zhng, C., Cai, M., ... Ma, J. (2015). The durability of carbon supported Pt nanowire as novel cathode catalyst for a 1.5 kW PEMFC stack. *Applied Catalysis B, 162*, 133–140.

Li, D., Lv, H., Kang, Y., Markovic, N. M., & Stamenkovic, V. R. (2016). Progress in the development of oxygen reduction reaction catalysts for low-temperature fuel cells. *Annual Review of Chemical and Biomolecular Engineering, 7*, 509–532.

Li, F.-M., Gao, X.-Q., Li, S.-N., Chen, Y., & Lee, J.-M. (2015). Thermal decomposition synthesis of functionalized PdPt alloy nanodendrites with high selectivity for oxygen reduction reaction. *NPG Asia Materials, 7*, e219.

Li, H.-H., Ma, S.-Y., Fu, Q.-Q., Liu, X.-J., Wu, L., & Yu, S.-H. (2015). Scalable bromide-triggered synthesis of Pd@Pt core–shell ultrathin nanowires with enhanced electrocatalytic performance toward oxygen reduction reaction. *Journal of the American Chemical Society, 137*, 7862–7868.

Li, H.-H., Xie, M.-L., Cui, C.-H., He, D., Gong, M., Jiang, J., ... Yu, S.-H. (2016). Surface charge polarization at the interface: Enhancing the oxygen reduction via precise synthesis of heterogeneous ultrathin Pt/PtTe nanowire. *Chemistry of Materials, 28*, 8890–8898.

Liang, Y.-T., Lin, S.-P., Liu, C.-W., Chung, S.-R., Chen, T.-Y., ... Wang, K.-W. (2015). The performance and stability of the oxygen reduction reaction on Pt-M (M = Pd, Ag and Au) nanorods: An experimental and computational study. *Chemical Communications, 51*, 6605–6608.

Liu, H., An, W., Li, Y., Frenkel, A. I., Sasaki, K., Koenigsmann, C., ... Wong, S. S. (2015). In situ probing of the active site geometry of ultrathin nanowires for the oxygen reduction reaction. *Journal of the American Chemical Society, 137*, 12597–12609.

Liu, H., Koenigsmann, C., Adzic, R. R., & Wong, S. S. (2014). Probing ultrathin one-dimensional Pd–Ni nanostructures as oxygen reduction reaction catalysts. *ACS Catalysis, 4*, 2544–2555.

Liu, J., Xu, C., Liu, C., Wang, F., Liu, H., Ji, J., & Li, Z. (2015). Impact of Cu-Pt nanotubes with a high degree of alloying on electro-catalytic activity toward oxygen reduction reaction. *Electrochimica Acta, 152*, 425–432.

Liu, L., Chen, L.-X., Wang, A.-J., Yuan, J., Shen, L., & Feng, J.-J. (2016). Hydrogen bubbles template-directed synthesis of self-supported AuPt nanowire networks for improved ethanol oxidation and oxygen reduction reactions. *International Journal of Hydrogen Energy, 41*, 8871–8880.

Liu, S., Zhang, Z., Bao, J., Lan, Y., Tu, W., Han, M., & Dai, Z. (2013). Controllable synthesis of tetragonal and cubic phase Cu_2Se nanowires assembled by small nanocubes and their electrocatalytic performance for oxygen reduction reaction. *Journal of Physical Chemistry C, 117*, 15164–15173.

Lu, Y., Du, S., & Steinberger-Wilckens, R. (2016a). One-dimensional nanostructured electrocatalysts for polymer electrolyte membrane fuel cells—A review. *Applied Catalysis B, 199*, 292–314.

Lu, Y., Du, S., & Steinberger-Wilckens, R. (2016b). Three-dimensional catalyst electrodes based on PtPd nanodendrites for oxygen reduction reaction in PEFC applications. *Applied Catalysis B, 187*, 108–114.

Lu, Y., Jiang, Y., & Chen, W. (2013). PtPd porous nanorods with enhanced electrocatalytic activity and durability for oxygen reduction reaction. *Nano Energy, 2*, 836–844.

Matanović, I., Kent, P. R. C., Garzon, F. H., & Henson, N. J. (2012). Density functional theory study of oxygen reduction activity on ultrathin platinum nanotubes. *Journal of Physical Chemistry C, 116*, 16499–16510.

Matanović, I., Kent, P. R. C., Garzon, F. H., & Henson, N. J. (2013). Density functional study of the structure, stability and oxygen reduction activity of ultrathin platinum nanowires. *Journal of Electrochemical Society, 160*, F548–F553.

Mazumder, V., Lee, Y., & Sun, S. (2010). Recent development of active nanoparticle catalysts for fuel cell reactions. *Advanced Functional Materials, 20*, 1224–1231.

Meng, H., Zhan, Y., Zeng, D., Zhang, X., Zhang, G., & Jaouen, F. (2015). Factors influencing the growth of Pt nanowires via chemical self-assembly and their fuel cell performance. *Small, 11*, 3377–3386.

Morozan, A., Jousselme, B., & Palacin, S. (2011). Low-platinum and platinum-free catalysts for the oxygen reduction reaction at fuel cell cathodes. *Energy & Environmental Science, 4*, 1238–1254.

Nie, Y., Li, L., & Wei, Z. (2015). Recent advancements in Pt and Pt-free catalysts for oxygen reduction reaction. *Chemical Society Reviews, 44*, 2168–2201.

Nørskov, J. K., Bligaard, T., Rossmeisl, J., & Christensen, C. H. (2009). Towards the computational design of solid catalysts. *Nature Chemistry, 1*, 37–46.

Nørskov, J. K., Rossmeisl, J., Logadottir, A., Lindqvist, L., Kitchin, J. R., Bligaard, T., & Jonsson, H. (2004). Origin of the overpotential for oxygen reduction at a fuel-cell cathode. *Journal of Physical Chemistry B, 108*, 17886–17892.

Peng, X., & Wong, S. S. (2009). Functional covalent chemistry of carbon nanotube surfaces. *Advanced Materials, 21*, 625–642.

Peng, X., Chen, J., Misewich, J. A., & Wong, S. S. (2009). Carbon nanotube-nanocrystal heterostructures. *Chemical Society Reviews, 38*, 1076–1098.

Rabis, A., Rodriguez, P., & Schmidt, T. J. (2012). Electrocatalysis for polymer electrolyte fuel cells: Recent achievements and future challenges. *ACS Catalysis, 2*, 864–890.

Ruan, L., Zhu, E., Chen, Y., Lin, Z., Huang, X., Duan, X., & Huang, Y. (2013). Biomimetic synthesis of an ultrathin platinum nanowire network with a high twin density for enhanced electrocatalytic activity and durability. *Angewandte Chemie International Edition, 52*, 12577–12581.

Shao, M., Chang, Q., Dodelet, J.-P., & Chenitz, R. (2016). Recent advances in electrocatalysts for oxygen reduction reaction. *Chemical Reviews, 116*, 3594–3657.

Shao, Y., Yin, G., & Gao, Y. (2007). Understanding and approaches for the durability issues of Pt-based catalysts for PEM fuel cell. *Journal of Power Sources, 171*, 558–566.

Shao-Horn, Y., Sheng, W., Chen, S., Ferreira, P., Holby, E., & Morgan, D. (2007). Instability of supported platinum nanoparticles in low-temperature fuel cells. *Topics in Catalysis, 46*, 285–305.

Su, K., Sui, S., Yao, X., Wei, Z., Zhang, J., & Du, S. (2014). Controlling Pt loading and carbon matrix thickness for a high performance Pt-nanowire catalyst layer in PEMFCs. *International Journal of Hydrogen Energy, 39*, 3397–3403.

Su, K., Yao, X., Sui, S., Wei, Z., Zhang, J., & Du, S. (2014). Ionomer content effects on the electrocatalyst layer with *in-situ* grown Pt nanowires in PEMFCs. *International Journal of Hydrogen Energy, 39*, 3219–3225.

Su, K., Yao, X., Sui, S., Wei, Z., Zhang, J., & Du, S. (2015). Matrix material study for in situ grown Pt nanowire electrocatalyst layer in proton exchange membrane fuel cells (PEMFCs). *Fuel Cells, 15*, 449–455.

Sung, M.-T., Chang, M.-H., & Ho, M.-H. (2014). Investigation of cathode electrocatalysts composed of electrospun Pt nanowires and Pt/C for proton exchange membrane fuel cells. *Journal of Power Sources, 249*, 320–326.

Tang, Y., & Cheng, W. (2015). Key parameters governing metallic nanoparticle electrocatalysis. *Nanoscale, 7*, 16151–16164.

Tuci, G., Zafferoni, C., D'Ambrosio, P., Caporali, S., Ceppatelli, M., Rossin, A., ... Giambastiani, G. (2013). Tailoring carbon nanotube n-dopants while designing metal-free electrocatalysts for the oxygen reduction reaction in alkaline medium. *ACS Catalysis, 3*, 2108–2111.

van Beurden, P., & Kramer, G. J. (2004). Atomistic mechanisms for the (1 x 1) to hex surface phase transformations of Pt(100). *Journal of Chemical Physics, 121*, 2317–2325.

Wang, C., Daimon, H., Onodera, T., Koda, T., & Sun, S. (2008). A general approach to the size and shape controlled synthesis of platinum nanoparticles and their catalytic reduction of oxygen. *Angewandte Chemie International Edition, 47*, 3588–3591.

Wang, J. X., Markovic, N. M., & Adzic, R. R. (2004). Kinetic analysis of oxygen reduction on Pt(111) in acid solutions: Intrinsic kinetic parameters and anion adsorption effects. *Journal of Physical Chemistry B, 108*, 4127–4133.

Wang, J. X., Zhang, J., & Adzic, R. R. (2007). Double-trap kinetic equation for the oxygen reduction reaction on Pt(111) in acidic media. *Journal of Physical Chemistry A, 111*, 12702–12710.

Wang, R., Higgins, D. C., Prabhudev, S., Lee, D. U., Choi, J.-Y., Hoque, M. A., ... Chen, Z. (2015). Synthesis and structural evolution of Pt nanotubular skeletons: Revealing the source of the instability of nanostructured electrocatalysts. *Journal of Materials Chemistry A, 3*, 12663–12671.

Wang, S., Jiang, S. P., Wang, X., & Guo, J. (2011). Enhanced electrochemical activity of Pt nanowire network electrocatalysts for methanol oxidation reaction of fuel cells. *Electrochimica Acta, 56*, 1563–1569.

Wang, W., Lei, B., & Guo, S. (2016). Engineering multimetallic nanocrystals for highly efficient oxygen reduction catalysts. *Advance Energy Materials, 6*, 1600236-n/a.

Wang, W., Lv, F., Lei, B., Wan, S., Luo, M., & Guo, S. (2016). Tuning nanowires and nanotubes for efficient fuel-cell electrocatalysis. *Advanced Materials, 28*, 10117–10141.

Watanabe, M., Tryk, D. A., Wakisaka, M., Yano, H., & Uchida, H. (2012). Overview of recent developments in oxygen reduction electrocatalysis. *Electrochimica Acta, 84*, 187–201.

Wittkopf, J. A., Zheng, J., & Yan, Y. (2014). High-performance dealloyed PtCu/CuNW oxygen reduction reaction catalyst for proton exchange membrane fuel cells. *ACS Catalysis, 4*, 3145–3151.

Wu, L., Liu, Z., Xu, M., Zhang, J., Yang, X., Huang, Y., . . . Tang, Y. (2016). Facile synthesis of ultrathin Pd—Pt alloy nanowires as highly active and durable catalysts for oxygen reduction reaction. *International Journal of Hydrogen Energy, 41*, 6805—6813.

Xiao, L., Zhuang, L., Liu, Y., Lu, J., & Abruña, Hc. D. (2008). Activating Pd by morphology tailoring for oxygen reduction. *Journal of the American Chemical Society, 131*, 602—608.

Xiao, Q., Cai, M., Balogh, M., Tessema, M., & Lu, Y. (2012). Symmetric growth of Pt ultrathin nanowires from dumbbell nuclei for use as oxygen reduction catalysts. *Nano Research, 5*, 145—151.

Xu, G.-R., Wang, B., Zhu, J.-Y., Liu, F.-Y., Chen, Y., Zeng, J.-H., . . . Lee, J.-M. (2016). Morphological and interfacial control of platinum nanostructures for electrocatalytic oxygen reduction. *ACS Catalysis, 6*, 5260—5267.

Yang, D., Yan, Z., Li, B., Higgins, D. C., Wang, J., Lv, H., . . . Zhang, C. (2016). Highly active and durable Pt—Co nanowire networks catalyst for the oxygen reduction reaction in PEMFCs. *International Journal of Hydrogen Energy, 41*, 18592—18601.

Yang, T., Cao, G., Huang, Q., Ma, Y., Wan, S., Zhao, H., . . . Yin, F. (2015). Surface-limited synthesis of Pt nanocluster decorated Pd hierarchical structures with enhanced electrocatalytic activity toward oxygen reduction reaction. *ACS Applied Materials & Interfaces, 7*, 17162—17170.

Yang, T., Ma, Y., Huang, Q., Cao, G., Wan, S., Zhao, H., . . . Yin, F. (2015). Palladium—iridium nanowires for enhancement of electro-catalytic activity towards oxygen reduction reaction. *Electrochemistry Communications, 59*, 95—99.

Yang, Y., Jin, H., Kim, H. Y., Yoon, J., Park, J., Baik, H., . . . Lee, K. (2016). Ternary dendritic nanowires as highly active and stable multifunctional electrocatalysts. *Nanoscale, 8*, 15167—15172.

Yang, Z., Zhou, X., Nie, H., Yao, Z., & Huang, S. (2011). *Facile construction of manganese oxide doped carbon nanotube catalysts with high activity for oxygen reduction reaction and investigations into the origin of their activity enhancement, ACS Applied Materials & Interfaces* (3, pp. 2601—2606).

Yeh, T.-H., Liu, C.-W., Chen, H.-S., & Wang, K.-W. (2013). Preparation of carbon-supported PtM (M = Au, Pd, or Cu) nanorods and their application in oxygen reduction reaction. *Electrochemistry Communications, 31*, 125—128.

You, H., Yang, S., Ding, B., & Yang, H. (2013). Synthesis of colloidal metal and metal alloy nanoparticles for electrochemical energy applications. *Chemical Society Reviews, 42*, 2880—2904.

Yu, F., Zhou, W., Bellabarba, R. M., & Tooze, R. P. (2014). One-step synthesis and shape-control of CuPd nanowire networks. *Nanoscale, 6*, 1093—1098.

Yu, X., & Ye, S. (2007). Recent advances in activity and durability enhancement of Pt/C catalytic cathode in PEMFC: Part II: Degradation mechanism and durability enhancement of carbon supported platinum catalyst. *Journal of Power Sources, 172*, 145—154.

Zhao, A., Masa, J., Schuhmann, W., & Xia, W. (2013). Activation and stabilization of nitrogen-doped carbon nanotubes as electrocatalysts in the oxygen reduction reaction at strongly alkaline conditions. *Journal of Physical Chemistry C, 117*, 24283—24291.

One-Dimensional Nanostructured Catalysts for Hydrocarbon Oxidation Reaction

Yaxiang Lu[1] and Shangfeng Du[2]

[1]Institute of Physics, Chinese Academy of Sciences, Beijing, China [2]School of Chemical Engineering, University of Birmingham, Birmingham, United Kingdom

In addition to using hydrogen for proton exchange membrane fuel cells (PEMFCs), liquid fuels typically in hydrocarbon species such as methanol, ethanol, and formic acid are also considered to be promising fuels as they can be handled, stored, and transported much easier. These hydrocarbons possess identical advantages such as abundant, inexpensive, and high volumetric energy density (Soloveichik, 2014; Zhang & Liu, 2009), at the same time, having their individual characteristics. Among these three fuels, methanol is widely investigated because it has a high energy density of $4.8\,\mathrm{kWh\,L^{-1}}$ (vs $2.4\,\mathrm{kWh\,L^{-1}}$ of liquid H_2) and a relative fast anodic reaction rate due to the cleavage of the $C-H$ bond. Compared with methanol, the oxidation of ethanol is associated with the breaking of $C-C$ bond, which requires a higher activation energy. Regardless of this, ethanol possesses a higher theoretical energy density of $6.3\,\mathrm{kWh\,L^{-1}}$ and can be easily obtained in large quantities from fermentation of biomass. Though formic acid has a lower energy density of $2.1\,\mathrm{kWh\,L^{-1}}$, recently, it draws a lot of attention for PEMFC applications due to its nontoxic and nonflammable property and can be oxidized at a less positive potential with faster kinetics than methanol at room temperature (Hong, Wang, & Wang, 2014; Lu, Du, & Steinberger-Wilckens, 2016).

To complete oxygen reduction reaction (ORR) at the cathode side in PEMFCs, these hydrocarbons need to conduct following reactions at the anode side, namely the methanol oxidation reaction (MOR) (Koenigsmann & Wong, 2011), ethanol oxidation reaction (EOR) (Friedl & Stimming, 2013), and formic acid oxidation reaction (FAOR) (MARKOVI, 1995), which are listed as in Eqs. (5.1−5.3):

One-Dimensional Nanostructures for PEM Fuel Cell Applications. DOI: http://dx.doi.org/10.1016/B978-0-12-811112-3.00005-4

$$CH_3OH + H_2O \rightarrow CO_2 + 6H^+ + 6e^- \qquad (5.1)$$

$$CH_3CH_2OH + 3H_2O \rightarrow 2CO_2 + 12H^+ + 12e^- \qquad (5.2)$$

$$HCOOH \rightarrow CO_2 + 2H^+ + 2e^- \qquad (5.3)$$

According to the above reaction equations, it can be assumed that the more the electrons are involved and the more the complicated reactions will proceed. Generally, the above reactions follow a multistep mechanism with a number of adsorbed reaction intermediates. Even the simplest FAOR has been reported to proceed via a "dual-path" mechanism with a "direct" dehydrogenation (Eq. 5.3) and "indirect" dehydration ($HCOOH \rightarrow CO + H_2O \rightarrow CO_2 + 2H^+ + 2e^-$) pathway (Xu et al., 2013). And the exact mechanism for the reaction process and the nature of the weakly adsorbed reaction intermediates (e.g., CHO, COOH, HCOO) are still under debate. Regarding MOR and EOR, mechanisms are even more complicated. Consequently, in fuel cells, hydrocarbon oxidations usually suffer from slow reaction kinetics and a high catalyst loading is required, e.g., 4 mg_{PtRu} cm^{-2} for MOR in the anode of direct methanol fuel cells versus 0.1 and 0.4 mg_{Pt} cm^{-2} for hydrogen oxidation reaction and ORR in the hydrogen PEMFC anode and cathode, respectively (Zhang & Liu, 2009). In order to efficiently employ these liquid fuels in fuel cells, catalysts with excellent catalytic activities and stability for hydrocarbon oxidation are highly desired.

By analogous to cathode ORR, electrocatalysts for anode hydrocarbon oxidation reactions are also based upon carbon supported Pt or Pt alloy nanoparticles initially (Soloveichik, 2014). However, apart from the heavily poisoned active sites by the strong adsorption of CO intermediates, the critical issue with these nanoparticulate catalysts is the poor stability. Comparing with zero-dimensional (0D) particles, one-dimensional (1D) catalysts, due to their inherent structural characteristics, usually show better tolerance to the CO intermediates and degradation. Further details of the mechanisms of the advantages of 1D catalysts can be found in Chapter 2, Advantages and Challenges of 1D Nanostructures for Fuel Cell Applications. Another unique advantage of 1D catalysts is that they can form highly porous catalyst layers. In liquid fuel cells, due to the large catalyst loading, the catalyst layer usually has more than 10 times of thickness than that in H_2 PEMFCs. The porous structure from 1D catalysts can facilitate mass exchange and fuel diffusion during the electrochemical process, leading to a higher catalyst utilization ratio. Therefore, 1D nanostructured catalysts have attracted huge amount of efforts for hydrocarbon oxidation research in recent years.

5.1 ONE-DIMENSIONAL Pt-BASED CATALYSTS FOR HYDROCARBON OXIDATION REACTION

Up to now, Pt-based alloys are still the most common catalysts for hydrocarbon oxidation reaction in fuel cells. However, the process of hydrocarbon oxidation usually undergoes formation of carbonaceous molecules such as CO and CHO, which poison the platinum surface and cause irreversible inactivation of catalysts. Compared with the much rough surface of nanoparticles, 1D Pt-based catalysts usually possess a smoother surface with less defect, which has a weak adsorption with those carbonaceous molecules (Huang, Sun, & Wang, 2011; Wang, Jiang, Wang, & Guo, 2011), thus providing an enhanced poisoning resistance. In addition to dimensionality, selectively control the exposed facet can also improve the performance of 1D Pt catalysts due to the minimized edges/corners. Moreover, to develop economically more affordable catalysts while enhancing the CO tolerance, alloy, or hybridize Pt with suitable less expensive metals and engineering the composition of Pt-based catalysts are also dedicated to further improve the hydrocarbon oxidation reaction.

5.1.1 One-Dimensional Pt Catalysts for Hydrocarbon Oxidation Reaction

Since Pt catalysts still play a critical role in a wide range of heterogeneous catalytic reactions, the controlled synthesis and fundamental investigation of factors that determine the catalytic properties are quite important. It has already been demonstrated that the low-energy {111} facets are the most active facet toward ORR, which also triggers the extensive efforts in the creation of particular facet to improve the kinetic of hydrocarbon oxidation reaction (Bu et al., 2016).

Polycrystalline and single-crystal Pt nanowires (NWs) (Ruan et al., 2013; Xia, Wu, Yan, Lou, & Wang, 2013), nanotubes (NTs) (Alia et al., 2010; Lou et al., 2016), nanofibers (NFs) (Wei et al., 2012), nanorods (NRs) (Li, Sato, & Yamauchi, 2013), and nanochain (NCs) (Shang, Hong, Guo, Wang, & Wang, 2016) have been tested for hydrocarbon oxidation with a main focus on MOR. For example, single-crystal Pt NWs synthesized by using solvothermal method were evaluated as catalysts for MOR and FAOR (Xia et al., 2013). The

NWs have a diameter of 3 nm and length of 10 μm (Fig. 5.1), showing mass activities of c. 500 and 700 mA mg^{-1} and specific activities of 1.15 and 1.5 mA cm^{-2} toward MOR and FAOR, respectively. In addition, after accelerated stability tests (AST) by 3000 potential cycles, they showed only about 31% and 37% activity loss for MOR and FAOR, which was much lower than the degradation rate of Pt/C.

The synergistic effect between catalyst and support for an improved catalytic performance was also further confirmed toward hydrocarbon oxidation reaction. Many conductive materials including carbon spheres (Meng, Xie, Chen, Sun, & Shen, 2011; Si, Ma, Liu, Zhang, & Xing, 2012), carbon NTs (Rajesh et al., 2013), graphene (Du, Lu, Malladi, Xu, & Steinberger-Wilckens, 2014; Sahu, Samantara, Satpati, Bhattacharjee, & Jena, 2013; Wang, Higgins, et al., 2013), and metal oxides (Ho et al., 2012) are employed to support 1D catalysts for

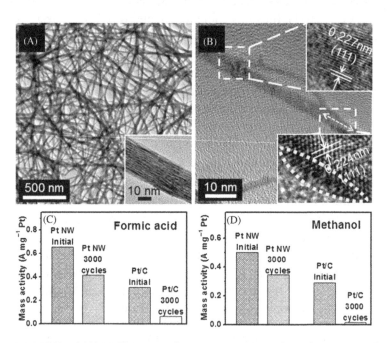

Figure 5.1 (A) TEM and (B) HRTEM images of Pt nanowire (NW) assemblies. The inset in (A) shows the corresponding high-magnification images. The surface atomic arrangements of NWs are also shown in the inset of (B), which indicates the existence of step structures on the surface of Pt NWs. The double arrows in (B) indicate the (111) growth direction of Pt NWs. Electrochemical evaluation: comparative mass activity for formic acid (C) and methanol oxidation (D) before and after accelerated stability tests of Pt NW membrane and commercial Pt/ C electrocatalysts. Adapted with permission from Xia, B.Y., Wu, H.B., Yan, Y., Lou, X.W., & Wang, X. (2013). Ultrathin and ultralong single-crystal platinum nanowire assemblies with highly stable electrocatalytic activity. Journal of the American Chemical Society, 135, 9480–9485. Copyright (2013) American Chemical Society.

MOR. Graphene-branched-Pt hybrid nanostructures (BPtNs) were synthesized by in situ reduction of graphene oxide and Pt precursor solution using $NaBH_4$ (Sahu et al., 2013), showing a peak current density of 49 and 14 times higher than that of BPtNs without graphene and Pt/C for MOR, respectively.

5.1.2 One-Dimensional Pt-Based Alloy Catalysts for Hydrocarbon Oxidation Reaction

The intermediates especially CO-like species generated during the hydrocarbon oxidation has a strong affinity to Pt, which can lead to a rapid poisoning of the catalyst surface, resulting in a fast performance degradation. Comparing with monometal systems, alloyed catalysts with controlled architectures and compositions seem to be an effective strategy. They can take advantages of the weaker CO adsorption energy of 1D Pt catalysts and/or the lower CO oxidation potential through adsorbed hydroxyl groups on non-Pt metals, which are denoted as electronic effects and bifunctional mechanism, respectively. Based on these understandings, numerous 1D Pt-based alloy nanostructures have been successfully synthesized and investigated for hydrocarbon oxidation reaction. Prominent examples include 1D PtPd (Koenigsmann & Wong, 2013; Lim, Jiang, Yu, Camargo, & Xia, 2010; Zhu, Guo, & Dong, 2012), PtAu (Kim et al., 2012) bimetallic nanocrystals and multimetallic nanoalloys based on them. In recent research, Rh (Shen, Gong, Xiao, & Wang, 2017) and Ru (Zheng et al., 2015) were also employed to construct 1D Pt-based alloy catalysts. Non-precious metals such as Fe, Co, Ni, Cu, Bi, and Mo have also been explored as substitutes for precious metals in 1D Pt-based electrocatalysts (Ding, Wang, et al., 2012; Du, Su, Wang, Frenkel, & Teng, 2011; Lu, Eid, et al., 2016; Yu, Wang, Peng, & Li, 2013). In addition to the chemical composition, the shape and morphology of alloyed nanostructures were also controlled to further improve the catalytic function and application performance.

It has been recently reported (Koenigsmann & Wong, 2013) that the incorporation of Pd with Pt can change the hydrocarbon oxidation process, leading to a CO-free pathway, which is different from the supposed bifunctional effect (e.g., PtRu alloy for MOR) to improve CO tolerance. The modification of the traditionally expected methanol oxidation pathway is ascribed to the decrease of Pt − Pt pair sites at the catalytic interface. Ding et al. (2014) showed a higher current density of PtPd porous hollow nanorod arrays (PHNRAs) toward MOR than Pt/C, as well as an outstanding stability confirmed by an almost constant peak current density after 500 potential cycles. The excellent ability of PtPd PHNRAs for CO antipoisoning was demonstrated by CO stripping. The catalysts can facilitate the removal of the adsorbed CO from their surfaces and also show higher CO oxidation ability than PtPd/C, PtPd film, and Pd/C catalysts. The performance improvement was ascribed to the alternative arrangement of Pt and Pd nanocrystals as well as the porous, hollow NR structure. They benefit for the modification of electronic structure and facilitate the mass transfer through the catalyst, finally leading to an efficient catalytic reaction.

Not like PtPd nanostructures, the combination of Au to Pt nanostructures mainly shows enhancement in the FAOR activity (Li, Meng, Wang, Jiang, & Zhu, 2016; Liu, Wei, Liu, & Wang, 2012). The alloyed PtAu changed the reaction pathway, enhancing the catalytic activity and the tolerance to CO poisoning through ensemble effects and modified electronics induced by compositional variation. The activity of the Pt_1Au_3 NTs for FAOR could be 26, 82, and 149 times higher than that of Pt NTs, Pt black, and Pt/C, respectively (Kim et al., 2012). In addition to Pd and Au, alloying other precious metals with Pt has also been reported. PtRu alloy NWs were prepared as high-performance catalyst for MOR (Li et al., 2012), attributing to the lattice contraction, the enhanced electronic property, along with the facilitated oxidation of adsorbed CO species. With the assist of Rh nanocubes as seeds, ultrathin Rh/Pt NWs have been synthesized and employed for EOR (Yuan, Zhou, Zhuang, & Wang, 2010). The Fourier transform infrared spectroscopy investigation for the reaction intermediates showed that Rh had a special effect on cleavage of C–C bond in ethanol and the alloyed catalyst possessed a high selectivity to the complete oxidation of ethanol to CO_2.

Bring a step forward from the precious metals, alloying Pt with less expensive transition metals can not only promote the formation of 1D morphology, but also increase the oxidation of intermediate species thus changing Pt d-band center, promoting C–H cleavage and the mitigation of CO oxidation for an improved hydrocarbon oxidation activity. 1D PtNi alloy nanostructures show excellent activity toward ORR, and it is also a promising candidate for hydrocarbon oxidation. PtNi nanodendrites (NDs) (Lee et al., 2013), NWs (Yu et al., 2013), NRs

(Qiu, Li, Lang, Zou, & Huang, 2012), and NTs (Ding, Wang, et al., 2012) have been investigated for FAOR, MOR, and EOR. It was also found that the addition of phosphorus (P) to PtNi significantly improved the relative content of Pt(0) and the 5d electron density of Pt for an enhanced catalytic activity. The specific current density of Pt-Ni-P nanotube arrays (NTAs) for MOR reaches 3.85 mA cm^{-2}. SEM image, scheme of MOR process with Pt-Ni-P NTAs and corresponding electrochemical activity and durability are shown in Fig. 5.2 (Ding, Wang, et al., 2012). This structure from Pt-Au-P was also evaluated for MOR very recently. It further proved that the addition of P led to a homogeneous distribution of nanocrystals and an increased the electrochemically active surface area (ECSA). The catalyst presents a more powerful ability of inhibiting CO formation to avoid CO poisoning during methanol electrooxidation (Zhang et al., 2017). Other transitional metal alloyed 1D Pt nanostructures recently studied include PtBi (Du et al., 2011) and worm-like Pt-M (M = Cu, Co, Fe) NWs (Yu et al., 2013) from wet-chemical approaches, as well as PtCo NWs

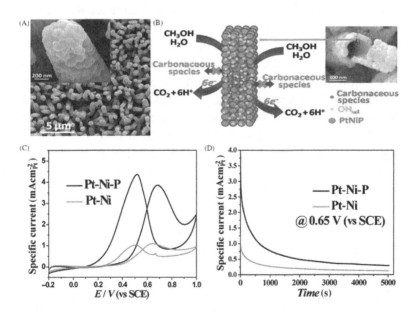

Figure 5.2 (A) SEM image of Pt-Ni-P nanotube arrays (NTAs); (B) scheme for the almost complete oxidation of carbonaceous species generated during methanol electrooxidation in the porous walls of Pt-Ni-P NTAs; (C) and (D) are CVs and chronoamperometry curves of Pt-Ni-P and Pt-Ni NTAs in 0.5 M CH₃OH + 0.5 M H₂SO₄ at 50 mV/s, respectively. Adapted with permission from Ding, L.X., Wang, A.L., Li, G.R., Liu, Z.Q., Zhao, W.X., Su, C.Y., & Tong, Y.X. (2012). Porous Pt-Ni-P composite nanotube arrays: Highly electroactive and durable catalysts for methanol electrooxidation. *Journal of the American Chemical Society, 134,* 5730−5733. Copyright (2012) American Chemical Society.

(Bertin, Garbarino, Ponrouch, & Guay, 2012), PtCu (Zhang, Li, Dong, Wang, & Webley, 2010), and PtCo (Luo, Yan, Xu, & Xue, 2013) NTs synthesized by template methods. All of them demonstrated increased catalytic activities and high CO poisoning resistance toward EOR or MOR.

The outstanding catalytic performance of PtPd and PtRu for MOR also makes them potential bases to develop multimetallic alloy catalysts to further improve the catalytic activities. Fe, Te, and Au alloyed PtPd NWs have been demonstrated showing a significant increase in the catalytic activity, a negative potential shift in methanol oxidation and a slow decay rate in durability test (Guo, Zhang, Sun, & Sun, 2011; Li, Zhao, et al., 2013). PtPdTe NWs with a diameter of 5−7 nm were synthesized exhibiting a high electrocatalytic activity of 595 mA cm^{-2} mg^{-1} toward MOR, which is 2.4- and 2.6-fold higher than that of PtTe NWs and Pt/C catalysts, respectively (Li, Zhao, et al., 2013). Furthermore, within the ternary alloy catalyst system, in addition to the bifunctional mechanism (e.g., associated with the alloying of Ru), the ligand effect can be further introduced (e.g., the presence of Fe in the alloy core) to lower the d-band center of Pt and thereby alter the electronic properties of the overall alloy as a stable and active catalyst. This research to PtRuM alloys with M = Ni, Co, W, and Fe was recently carefully studied by Wong's group (Koenigsmann & Wong, 2011; Scofield, Koenigsmann, Wang, Liu, & Wong, 2015) and excellent catalytic performance was confirmed. Moreover, they also found that the CO tolerance of the ternary alloy catalysts was not necessarily correlated with their corresponding MOR activity, as shown in Fig. 5.3. Pt$_7$Ru$_3$ NW catalyst possesses a better CO tolerance as compared with the Pt$_7$Ru$_2$Fe NW catalyst but maintains a lower MOR activity. A further study is required to clarify the fundamental mechanisms behind this.

5.1.3 One-Dimensional Pt-Based Hybrid Catalysts for Hydrocarbon Oxidation Reaction

The primary idea of employing hybrid structure is to integrate the 1D morphology, core/shell structure, alloy effect and/or hollow structure, etc. to combine the advantages of each part and further enhance the activity and stability toward hydrocarbon oxidation reaction.

Similar to the advantages for ORR as described in Chapter 4, One-Dimensional Nanostructured Catalysts for Oxygen Reduction

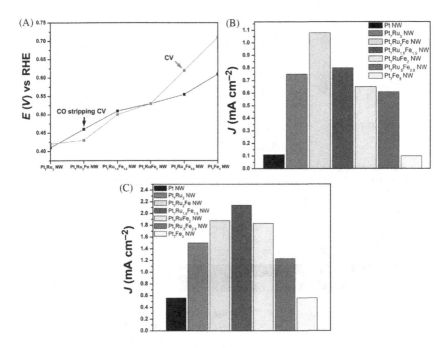

Figure 5.3 (A) A plot investigating the trend in onset potential for CO stripping (black) and the corresponding onset of surface oxide reduction as a function of systematically varying chemical composition from Pt_7Ru_3 NWs to Pt_7Fe_3 NWs. Bar graph highlighting (B) MOR and (C) FAOR activity at E (V) versus RHE = 0.65 V for Pt NWs, Pt_7Ru_3 NWs, Pt_7Ru_2Fe NWs, $Pt_7Ru_{1.5}Fe_{1.5}$ NWs, Pt_7RuFe_2 NWs, $Pt_7Ru._5Fe_{2.5}$ NWs, and Pt_7Fe_3 NWs. Adapted with permission from Scofield, M.E., Koenigsmann, C., Wang, L., Liu, H.Q., & Wong, S.S. (2015). Tailoring the composition of ultrathin, ternary alloy PtRuFe nanowires for the methanol oxidation reaction and formic acid oxidation reaction. Energy & Environmental Science, 8, 350–363. Copyright (2015) Royal Society of Chemistry.

Reaction, here 1D core-shell catalysts with a core of non-Pt precious metal (e.g., Au) (Lee et al., 2012) and transitional metals (e.g., Ni) (Ding, Li, et al., 2012), and a Pt shell are still the most common structures investigated for hydrocarbon oxidation. Taking the unique advantages of CeO_2 and conducting polyaniline (PANI) polymer, in particular their excellent ability of CO antipoisoning, Xu et al. prepared novel Pt/CeO_2/PANI three-layered hybrid hollow NR arrays (THNRAs) (Xu, Wang, Tong, & Li, 2016) by electrodeposition method using ZnO templates. Details of the architecture are shown in Fig. 5.4. The electron delocalization among Pt, CeO_2, and PANI leads to a high content of metallic Pt and synergistic effects for methanol oxidation. Moreover, the catalysts have the advantage of CeO_2 that can provide the labile OH species for electrooxidation of surface adsorbed carbonaceous intermediates, leading to high CO tolerance of catalysts. Going one step further, the surface Pt can be alloyed with other

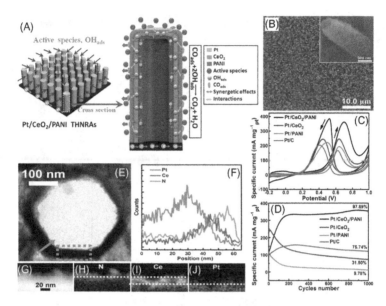

Figure 5.4 (A) Schematic illustration and (B) SEM image of Pt/CeO₂/PANI THNRAs; (C) cyclic voltammo-grams and (D) change in the specific peak current density after 1000 cycles of Pt/CeO₂/PANI THNRAs, Pt/PANI HNRAs, Pt/CeO₂ HNRAs, and commercial Pt/C catalysts in the solution of 0.5 M H₂SO₄ + 0.5 M CH₃OH at 100 mV s⁻¹; (E) TEM image of a frontal Pt/CeO₂/PANI hollow nanorod; (F) EDX line scans along the red arrow (dark gray in print versions) in Panel (E); (G) STEM-HAADF image, and (H-J) STEM-EDX-mappings of a part of wall of the nanorod marked with a green frame (light gray in print versions) in Panel (E). Adapted with permission from Xu, H., Wang, A.L., Tong, Y.X., & Li, G.R. (2016). Enhanced catalytic activity and stability of Pt/CeO₂/PANI hybrid hollow nanorod arrays for methanol electro-oxidation. ACS Catalysis, 6, 5198–5206. Copyright (2016) American Chemical Society.

metals to prepare 1D catalysts with core-alloy shell structures. The alloy shell also possesses the unique advantages of the multiple structure features and synergetic effects among different metals for hydrocarbon oxidation. Pt alloy that shells with metals such as Au (Liu et al., 2012), Pb (Zhang, Guo, Zhu, Guo, & Huang, 2016), and Co (Sriphathoorat et al., 2016) has been prepared, and enhanced activity and stability were also demonstrated.

Apart from the core-shell structure, Pt-based heteronanostructures were also demonstrated providing enhanced activities and stability toward hydrocarbon oxidation. Pt-on-Pd bimetallic NDs (Wang, Nemoto, & Yamauchi, 2011), Pt-decorated coral-like Pd NCs (Lan et al., 2013), Pt-on-Pd NWs (Huang et al., 2013), and Pt-decorated Ru NWs (Koenigsmann, Semple, Sutter, Tobierre, & Wong, 2013) have been reported for catalyzing hydrocarbon oxidation reaction either in acid or alkaline medium. Recently, some Pt-based ternary hybrid structures have also captured attention. $Pt-MoO_3-RGO$ ternary hybrid hollow NRs (Wang, Liang, Lu, Tong, & Li, 2016) with rich heterogeneous interfaces and strong electron interactions among Pt 4f orbitals, Mo 3d orbitals and RGO π-conjugated ligands demonstrated improved catalytic activity and durability for MOR.

5.2 ONE-DIMENSIONAL NON-Pt BASED CATALYSTS FOR HYDROCARBON OXIDATION REACTION

The high cost, and the tendency of the surface poisoning of Pt by the strong adsorption of CO intermediates in hydrocarbon oxidation force both academic and industrial communities to find Pt substitutes. In recent years, this research topic of 1D catalysts mainly focused on Pd and Pd-based nanostructures due to the great abundance of Pd in nature. Particularly, unlike its counterpart Pt, 1D Pd-based catalysts are less suffered from poisonous CO intermediate species and have a lower oxidation overpotential toward hydrocarbon oxidation (Hong, Wang, & Wang, 2014; Meng, Wang, Shen, & Wu, 2011), thus regarding as one of the most promising Pt-free catalysts for the electrooxidation of hydrocarbon molecules, but still with a focus in alkaline condition. The only 1D Pd free catalysts reported are Ni − Cu alloy porous NWs prepared for MOR also in alkaline solution (Ding et al., 2011).

Pd NWs (Hong, Wang, & Wang, 2014; Wang, Choi, et al., 2014), NCs (Zheng et al., 2014), nanothorns (Meng, Wang, et al., 2011), and more complex flower-like nanostructured networks (Ren et al., 2014), as well as Pd/polyaniline/Pd sandwich structured NTAs (Wang, Xu, et al., 2013) have been studied and they also showed better performance than related 0D nanoparticles. Xia et al. (Wang, Choi, et al., 2014) synthesized 2 nm ultrathin Pd NWs via a polyol method and the electrochemical measurement showed a catalytic current density of 2.5-folds higher than that of Pd/C catalysts toward FAOR. Hong, Wang, et al. (2014) and Wang, Chen, Liu, Li, & Sun (2010) reported that the mass activity of Pd NWs with a diameter of 4−5 nm reached 1.45 and 1.1 A mg^{-1} for EOR and FAOR, respectively.

The main efforts for 1D Pd-based alloy catalysts toward hydrocarbon oxidation were precious alloyed catalysts, especially 1D Pd-Au bimetallic nanocrystals. Pd-Au alloy NDs (Lee et al., 2010; Shi et al., 2013), Au@Pd core-shell NDs (Wang, Sun, Yang, & Su, 2013), and PdAu NW networks (Hong, Wang, et al., 2014) were all synthesized through the wet-chemical reduction method by using different reductants and stabilizing agents. By controlling the nucleation and growth rate, the produced catalysts with the diameter ranges between 3 and 26 nm demonstrated good catalytic performance for MOR or EOR in alkaline media. The excellent performance can be ascribed to the improved electron transport characteristics, the increased active sites as well as the favored adsorption of OH_{ads} onto the catalyst surface, which collectively improve the catalytic process. Furthermore, for FAOR, PdAg alloy NWs with an average diameter of 5−8 nm were obtained by coreduction method (Lu & Chen, 2011) and nanoneedle-covered PdAg NTs (Lu & Chen, 2010) were synthesized via a galvanic displacement of Ag NRs. Rh-on-Pd bimetallic NDs composed of Pd cores and 15−30 nm Rh branches were also synthesized with ethylene glycol and CTAB for EOR application (Shen & Zhao, 2013). Apart from combining with precious metals, some works were also reported for 1D Pd alloyed with nonprecious metals. Nanoporous PdNi (Du, Chen, Wang, & Yin, 2010) and PdBi (Liao, Zhu, & Hou, 2014) alloy NWs were synthesized for FAOR application. Quinary PdNiCoCuFe alloy NTAs have also been prepared by template-assisted electrodeposition method for MOR and EOR in alkaline solution (Wang, Wan, Xu, Tong, & Li, 2014).

5.3 SUMMARY AND OUTLOOK

The unique surface properties of 1D nanostructures endow them with excellent capability to tolerate and even hinder the formation of CO-like species, making them promising catalysts for hydrocarbon oxidation applications. Controlled synthesis of 1D catalysts to expose preferential facets, alloying with other metals to adjust the adsorption of hydrocarbon species and desorption of reaction intermediates, together with integrating novel structure, morphology, and architectures to fabricate hybrid catalysts have been the main strategies to realize the goal in the past 5 years. 1D Pt-based nanostructures still dominate the highly active and durable catalysts for hydrocarbon oxidation reaction. Some attentions have also been paid to Pt alternative catalysts especially Pd-based catalyst category, although most of them are still limited to hydrocarbon oxidation in the alkaline electrolyte.

Looking into the development of 1D catalysts for hydrocarbon oxidation, similar to that in ORR, the transition from ex situ half-cell measurement to the in situ test in membrane electrode assembly (MEA) in fuel cells is urgent for practical applications in future. Considering the unique advantage of 1D catalysts in forming porous catalyst layers, in particular the benefit in facilitating mass transfer through the thick catalyst layer in liquid fuel cells, 1D catalysts are expected to play a significant role for this kind of applications. Moreover, with the increasingly available infrastructure of ethanol and its environmental friendly feature, the research of 1D catalysts to facilitate the C$-$C cleavage for EOR in direct ethanol fuel cells is required to meet the increasing commercial potential.

REFERENCES

Alia, S. M., Zhang, G., Kisailus, D., Li, D., Gu, S., Jensen, K., & Yan, Y. (2010). Porous platinum nanotubes for oxygen reduction and methanol oxidation reactions. *Advanced Functional Materials, 20*, 3742–3746.

Bertin, E., Garbarino, S., Ponrouch, A., & Guay, D. (2012). Synthesis and characterization of PtCo nanowires for the electro-oxidation of methanol. *Journal of Power Sources, 206*, 20–28.

Bu, L. Z., Feng, Y. G., Yao, J. L., Guo, S. J., Guo, J., & Huang, X. Q. (2016). Facet and dimensionality control of Pt nanostructures for efficient oxygen reduction and methanol oxidation electrocatalysts. *Nano Research, 9*, 2811–2821.

Ding, L. X., Li, G. R., Wang, Z. L., Liu, Z. Q., Liu, H., & Tong, Y. X. (2012). Porous Ni@Pt core-shell nanotube array electrocatalyst with high activity and stability for methanol oxidation. *Chemistry − A European Journal, 18*, 8386–8391.

Ding, L. X., Liang, C. L., Xu, H., Wang, A. L., Tong, Y. X., & Li, G. R. (2014). Porous hollow nanorod arrays composed of alternating Pt and Pd nanocrystals with superior electrocatalytic activity and durability for methanol oxidation. *Advanced Materials Interfaces*, 1400005.

Ding, L. X., Wang, A. L., Li, G. R., Liu, Z. Q., Zhao, W. X., Su, C. Y., & Tong, Y. X. (2012). Porous Pt-Ni-P composite nanotube arrays: Highly electroactive and durable catalysts for methanol electrooxidation. *Journal of the American Chemical Society, 134*, 5730–5733.

Ding, R. M., Liu, J. P., Jiang, J., Wu, F., Zhu, J. H., & Huang, X. T. (2011). Tailored Ni-Cu alloy hierarchical porous nanowire as a potential efficient catalyst for DMFCs. *Catalysis Science & Technology, 1*, 1406–1411.

Du, C., Chen, M., Wang, W., & Yin, G. (2010). Nanoporous PdNi alloy nanowires as highly active catalysts for the electro-oxidation of formic acid. *ACS Applied Materials & Interfaces, 3*, 105–109.

Du, S., Lu, Y., Malladi, S. K., Xu, Q., & Steinberger-Wilckens, R. (2014). A simple approach for PtNi-MWCNT hybrid nanostructures as high performance electrocatalysts for the oxygen reduction reaction. *Journal of Materials Chemistry A, 2*, 692–698.

Du, W., Su, D., Wang, Q., Frenkel, A. I., & Teng, X. (2011). Promotional effects of bismuth on the formation of platinum − bismuth nanowires network and the electrocatalytic activity toward ethanol oxidation. *Crystal Growth & Design, 11*, 594−599.

Friedl, J., & Stimming, U. (2013). Model catalyst studies on hydrogen and ethanol oxidation for fuel cells. *Electrochimica Acta, 101*, 41−58.

Guo, S., Zhang, S., Sun, X., & Sun, S. (2011). Synthesis of ultrathin FePtPd nanowires and their use as catalysts for methanol oxidation reaction. *Journal of the American Chemical Society, 133*, 15354−15357.

Ho, V. T. T., Nguyen, N. G., Pan, C. J., Cheng, J. H., Rick, J., Su, W. N., ... Hwang, B. J. (2012). Advanced nanoelectrocatalyst for methanol oxidation and oxygen reduction reaction, fabricated as one-dimensional pt nanowires on nanostructured robust $Ti_{0.7}Ru_{0.3}O_2$ support. *Nano Energy, 1*, 687−695.

Hong, W., Wang, J., & Wang, E. (2014). Bromide ion mediated synthesis of carbon supported ultrathin palladium nanowires with enhanced catalytic activity toward formic acid/ethanol electrooxidation. *International Journal of Hydrogen Energy, 39*, 3226−3230.

Hong, W., Wang, J., & Wang, E. (2014). Facile synthesis of highly active PdAu nanowire networks as self-supported electrocatalyst for ethanol electrooxidation. *ACS Applied Materials & Interfaces, 6*, 9481−9487.

Hong, W., Wang, J., & Wang, E. K. (2014). Facile synthesis of highly active PdAu nanowire networks as self-supported electrocatalyst for ethanol electrooxidation. *ACS Applied Materials & Interfaces, 6*, 9481−9487.

Huang, H., Sun, D., & Wang, X. (2011). Low-defect MWNT−Pt nanocomposite as a high performance electrocatalyst for direct methanol fuel cells. *The Journal of Physical Chemistry C, 115*, 19405−19412.

Huang, Z. Y., Zhou, H. H., Chang, Y. W., Fu, C. P., Zeng, F. Y., & Kuang, Y. F. (2013). Improved catalytic performance of Pd nanowires for ethanol oxidation by monolayer of Pt. *Chemical Physics Letters, 585*, 128−132.

Kim, Y., Kim, H. J., Kim, Y. S., Choi, S. M., Seo, M. H., & Kim, W. B. (2012). Shape- and composition-sensitive activity of Pt and PtAu catalysts for formic acid electrooxidation. *The Journal of Physical Chemistry C, 116*, 18093−18100.

Koenigsmann, C., Semple, D. B., Sutter, E., Tobierre, S. E., & Wong, S. S. (2013). Ambient synthesis of high-quality ruthenium nanowires and the morphology-dependent electrocatalytic performance of platinum-decorated ruthenium nanowires and nanoparticles in the methanol oxidation reaction. *ACS Applied Materials & Interfaces, 5*, 5518−5530.

Koenigsmann, C., & Wong, S. S. (2011). One-dimensional noble metal electrocatalysts: A promising structural paradigm for direct methanol fuel cells. *Energy & Environmental Science, 4*, 1161−1176.

Koenigsmann, C., & Wong, S. S. (2013). Tailoring chemical composition to achieve enhanced methanol oxidation reaction and methanol-tolerant oxygen reduction reaction performance in palladium-based nanowire systems. *ACS Catalysis, 3*, 2031−2040.

Lan, F., Wang, D. L., Lu, S. F., Zhang, J., Liang, D. W., Peng, S. K., ... Xiang, Y. (2013). Ultra-low loading Pt decorated coral-like Pd nanochain networks with enhanced activity and stability towards formic acid electrooxidation. *Journal of Materials Chemistry A, 1*, 1548−1552.

Lee, Y., Kim, J., Yun, D. S., Nam, Y. S., Shao-Horn, Y., & Belcher, A. M. (2012). Virus-templated Au and Au-Pt core-shell nanowires and their electrocatalytic activities for fuel cell applications. *Energy & Environmental Science, 5*, 8328−8334.

Lee, Y. W., Kim, B. Y., Lee, K. H., Song, W. J., Cao, G. Z., & Park, K. W. (2013). Synthesis of monodispersed Pt-Ni alloy nanodendrites and their electrochemical properties. *International Journal of Electrochemcial Science, 8*, 2305−2312.

Lee, Y. W., Kim, M., Kim, Y., Kang, S. W., Lee, J.-H., & Han, S. W. (2010). Synthesis and electrocatalytic activity of Au − Pd alloy nanodendrites for ethanol oxidation. *The Journal of Physical Chemistry C*, *114*, 7689−7693.

Li, B., Higgins, D. C., Zhu, S. M., Li, H., Wang, H. J., Ma, J. X., & Chen, Z. W. (2012). Highly active Pt-Ru nanowire network catalysts for the methanol oxidation reaction. *Catalysis Communications*, *18*, 51−54.

Li, C., Sato, T., & Yamauchi, Y. (2013). Electrochemical synthesis of one-dimensional mesoporous Pt nanorods using the assembly of surfactant micelles in confined space. *Angewandte Chemie International Edition*, *52*, 8050−8053.

Li, D., Meng, F., Wang, H., Jiang, X., & Zhu, Y. (2016). Nanoporous AuPt alloy with low Pt content: A remarkable electrocatalyst with enhanced activity towards formic acid electrooxidation. *Electrochimica Acta*, *190*, 852−861.

Li, H. H., Zhao, S., Gong, M., Cui, C. H., He, D., Wu, L., ... Yu, S. H. (2013). Ultrathin PtPdTe nanowires as superior catalysts for methanol electrooxidation. *Angewandte Chemie International Edition*, *52*, 7472−7476.

Liao, H. B., Zhu, J. H., & Hou, Y. L. (2014). Synthesis and electrocatalytic properties of PtBi nanoplatelets and PdBi nanowires. *Nanoscale*, *6*, 1049−1055.

Lim, B., Jiang, M., Yu, T., Camargo, P. C., & Xia, Y. (2010). Nucleation and growth mechanisms for Pd-Pt bimetallic nanodendrites and their electrocatalytic properties. *Nano Research*, *3*, 69−80.

Liu, C. W., Wei, Y. C., Liu, C. C., & Wang, K. W. (2012). Pt-Au core/shell nanorods: Preparation and applications as electrocatalysts for fuel cells. *Journal of Materials Chemistry*, *22*, 4641−4644.

Lou, Y., Li, C. G., Gao, X. D., Bai, T. Y., Chen, C. L., Liang, C., ... Feng, S. H. (2016). Porous Pt nanotubes with high methanol oxidation electrocatalytic activity based on original bamboo-shaped Te nanotubes. *ACS Applied Materials & Interfaces*, *8*, 16147−16153.

Lu, S. L., Eid, K., Lin, M., Wang, L., Wang, H. J., & Gu, H. W. (2016). Hydrogen gas-assisted synthesis of worm-like PtMo wavy nanowires as efficient catalysts for the methanol oxidation reaction. *Journal of Materials Chemistry A*, *4*, 10508−10513.

Lu, Y., & Chen, W. (2010). Nanoneedle-covered Pd − Ag nanotubes: High electrocatalytic activity for formic acid oxidation. *The Journal of Physical Chemistry C*, *114*, 21190−21200.

Lu, Y., & Chen, W. (2011). PdAg alloy nanowires: Facile one-step synthesis and high electrocatalytic activity for formic acid oxidation. *ACS Catalysis*, *2*, 84−90.

Lu, Y. X., Du, S. F., & Steinberger-Wilckens, R. (2016). One-dimensional nanostructured electrocatalysts for polymer electrolyte membrane fuel cells-A review. *Applied Catalysis B-Environmental*, *199*, 292−314.

Luo, B. M., Yan, X. B., Xu, S., & Xue, Q. J. (2013). Synthesis of worm-like PtCo nanotubes for methanol oxidation. *Electrochemistry Communications*, *30*, 71−74.

MARKOVI, N. M. (1995). Electro-oxidation mechanisms on methanol and formic cid on Pt-Ru alloy surface. *Electrochimica Acta*, 91−98.

Meng, H., Wang, C. X., Shen, P. K., & Wu, G. (2011). Palladium thorn clusters as catalysts for electrooxidation of formic acid. *Energy & Environmental Science*, *4*, 1522−1526.

Meng, H., Xie, F. Y., Chen, J., Sun, S. H., & Shen, P. K. (2011). Morphology controllable growth of Pt nanoparticles/nanowires on carbon powders and its application as novel electrocatalyst for methanol oxidation. *Nanoscale*, *3*, 5041−5048.

Qiu, H., Li, L., Lang, Q., Zou, F., & Huang, X. (2012). Retracted article: Aligned nanoporous PtNi nanorod-like structures for electrocatalysis and biosensing. *RSC Advances*, *2*, 3548.

Rajesh, Paul, R. K., & Mulchandani, A. (2013). Platinum nanoflowers decorated three-dimensional graphene-carbon nanotubes hybrid with enhanced electrocatalytic activity. *Journal of Power Sources*, *223*, 23−29.

Ren, M., Zou, L., Yuan, T., Huang, Q., Zou, Z., Li, X., & Yang, H. (2014). Novel palladium flower-like nanostructured networks for electrocatalytic oxidation of formic acid. *Journal of Power Sources, 267*, 527–532.

Ruan, L. Y., Zhu, E. B., Chen, Y., Lin, Z. Y., Huang, X. Q., Duan, X. F. ,, ... Huang, Y. (2013). Biomimetic synthesis of an ultrathin platinum nanowire network with a high twin density for enhanced electrocatalytic activity and durability. *Angewandte Chemie-International Edition, 52*, 12577–12581.

Sahu, S. C., Samantara, A. K., Satpati, B., Bhattacharjee, S., & Jena, B. K. (2013). A facile approach for in situ synthesis of graphene-branched-Pt hybrid nanostructures with excellent electrochemical performance. *Nanoscale, 5*, 11265–11274.

Scofield, M. E., Koenigsmann, C., Wang, L., Liu, H. Q., & Wong, S. S. (2015). Tailoring the composition of ultrathin, ternary alloy PtRuFe nanowires for the methanol oxidation reaction and formic acid oxidation reaction. *Energy & Environmental Science, 8*, 350–363.

Shang, C. S., Hong, W., Guo, Y. X., Wang, J., & Wang, E. K. (2016). One-step synthesis of platinum nanochain networks toward methanol electrooxidation. *Chemelectrochem, 3*, 2093–2099.

Shen, S., & Zhao, T. (2013). One-step polyol synthesis of Rh-on-Pd bimetallic nanodendrites and their electrocatalytic properties for ethanol oxidation in alkaline media. *Journal of Materials Chemistry A, 1*, 906–912.

Shen, Y., Gong, B., Xiao, K. J., & Wang, L. (2017). In situ assembly of ultrathin PtRh nanowires to graphene nanosheets as highly efficient electrocatalysts for the oxidation of ethanol. *ACS Applied Materials & Interfaces, 9*, 3535–3543.

Shi, L., Wang, A., Zhang, T., Zhang, B., Su, D., Li, H., ... Song, Y. (2013). One-step synthesis of Au–Pd alloy nanodendrites and their catalytic activity. *The Journal of Physical Chemistry C, 117*, 12526–12536.

Si, F., Ma, L., Liu, C., Zhang, X., & Xing, W. (2012). The role of anisotropic structure and its aspect ratio: High-loading carbon nanospheres supported Pt nanowires with high performance toward methanol electrooxidation. *RSC Advances, 2*, 401–403.

Soloveichik, G. L. (2014). Liquid fuel cells. *Beilstein Journal of Nanotechnology, 5*, 1399–1418.

Sriphathoorat, R., Wang, K., Luo, S. P., Tang, M., Du, H. Y., Du, X. W., ... Shen, P. K. (2016). Well-defined PtNiCo core-shell nanodendrites with enhanced catalytic performance for methanol oxidation. *Journal of Materials Chemistry A, 4*, 18015–18021.

Wang, A. L., Liang, C. L., Lu, X. F., Tong, Y. X., & Li, G. R. (2016). Pt-MoO$_3$-RGO ternary hybrid hollow nanorod arrays as high-performance catalysts for methanol electrooxidation. *Journal of Materials Chemistry A, 4*, 1923–1930.

Wang, A.-L., Wan, H.-C., Xu, H., Tong, Y.-X., & Li, G.-R. (2014). Quinary PdNiCoCuFe alloy nanotube arrays as efficient electrocatalysts for methanol oxidation. *Electrochimica Acta, 127*, 448–453.

Wang, A.-L., Xu, H., Feng, J.-X., Ding, L.-X., Tong, Y.-X., & Li, G.-R. (2013). Design of Pd/PANI/Pd sandwich-structured nanotube array catalysts with special shape effects and synergistic effects for ethanol electrooxidation. *Journal of the American Chemical Society, 135*, 10703–10709.

Wang, H., Sun, Z., Yang, Y., & Su, D. (2013). The growth and enhanced catalytic performance of Au@Pd core-shell nanodendrites. *Nanoscale, 5*, 139–142.

Wang, J., Chen, Y., Liu, H., Li, R., & Sun, X. (2010). Synthesis of Pd nanowire networks by a simple template-free and surfactant-free method and their application in formic acid electrooxidation. *Electrochemistry Communications, 12*, 219–222.

Wang, L., Nemoto, Y., & Yamauchi, Y. (2011). Direct synthesis of spatially-controlled Pt-on-Pd bimetallic nanodendrites with superior electrocatalytic activity. *Journal of the American Chemical Society, 133*, 9674–9677.

Wang, R., Higgins, D. C., Hoque, M. A., Lee, D., Hassan, F., & Chen, Z. (2013). Controlled growth of platinum nanowire arrays on sulfur doped graphene as high performance electrocatalyst. *Sci. Rep.*, *3*, 2341.

Wang, S., Jiang, S. P., Wang, X., & Guo, J. (2011). Enhanced electrochemical activity of Pt nanowire network electrocatalysts for methanol oxidation reaction of fuel cells. *Electrochimica Acta*, *56*, 1563–1569.

Wang, Y., Choi, S.-I., Zhao, X., Xie, S., Peng, H.-C., Chi, M., ... Xia, Y. (2014). Polyol synthesis of ultrathin Pd nanowires via attachment-based growth and their enhanced activity towards formic acid oxidation. *Advanced Functional Materials*, *24*, 131–139.

Wei, L., Fan, Y. J., Wang, H. H., Tian, N., Zhou, Z. Y., & Sun, S. G. (2012). Electrochemically shape-controlled synthesis in deep eutectic solvents of Pt nanoflowers with enhanced activity for ethanol oxidation. *Electrochimica Acta*, *76*, 468–474.

Xia, B. Y., Wu, H. B., Yan, Y., Lou, X. W., & Wang, X. (2013). Ultrathin and ultralong single-crystal platinum nanowire assemblies with highly stable electrocatalytic activity. *Journal of the American Chemical Society*, *135*, 9480–9485.

Xu, H., Wang, A. L., Tong, Y. X., & Li, G. R. (2016). Enhanced catalytic activity and stability of $Pt/CeO_2/PANI$ hybrid hollow nanorod arrays for methanol electro-oxidation. *ACS Catalysis*, *6*, 5198–5206.

Xu, J., Yuan, D. F., Yang, F., Mei, D., Zhang, Z. B., & Chen, Y. X. (2013). On the mechanism of the direct pathway for formic acid oxidation at a Pt(111) electrode. *Physical Chemistry Chemical Physics*, *15*, 4367–4376.

Yu, X. F., Wang, D. S., Peng, Q., & Li, Y. D. (2013). Pt-M (M = Cu, Co, Ni, Fe) nanocrystals: From small nanoparticles to wormlike nanowires by oriented attachment. *Chemistry-a European Journal*, *19*, 233–239.

Yuan, Q., Zhou, Z., Zhuang, J., & Wang, X. (2010). Seed displacement, epitaxial synthesis of Rh/Pt bimetallic ultrathin nanowires for highly selective oxidizing ethanol to CO_2. *Chemistry of Materials*, *22*, 2395–2402.

Zhang, J., & Liu, H. (2009). *Electrocatalysis of direct methanol fuel cells: From fundamentals to applications*. Weinheim, Germany Chichester: Wiley-VCH, John Wiley distributor.

Zhang, L., Ding, L. X., Chen, H., Li, D., Wang, S., & Wang, H. (2017). Self-supported PtAuP alloy nanotube arrays with enhanced activity and stability for methanol electro-oxidation. *Small*, *13*, 1604000.

Zhang, N., Guo, S., Zhu, X., Guo, J., & Huang, X. Q. (2016). Hierarchical Pt/PtxPb core/shell nanowires as efficient catalysts for electrooxidation of liquid fuels. *Chemistry of Materials*, *28*, 4447–4452.

Zhang, X. Y., Li, D., Dong, D. H., Wang, H. T., & Webley, P. A. (2010). One-step fabrication of ordered Pt-Cu alloy nanotube arrays for ethanol electrooxidation. *Materials Letters*, *64*, 1169–1172.

Zheng, J., Cullen, D. A., Forest, R. V., Wittkopft, J. A., Zhuang, Z. B., Sheng, W. C., ... Yan, Y. S. (2015). Platinum-ruthenium nanotubes and platinum-ruthenium coated copper nanowires as efficient catalysts for electro-oxidation of methanol. *ACS Catalysis*, *5*, 1468–1474.

Zheng, J. N., Zhang, M., Li, F. F., Li, S. S., Wang, A. J., & Feng, J. J. (2014). Facile synthesis of Pd nanochains with enhanced electrocatalytic performance for formic acid oxidation. *Electrochimica Acta*, *130*, 446–452.

Zhu, C., Guo, S., & Dong, S. (2012). PdM (M = Pt, Au) bimetallic alloy nanowires with enhanced electrocatalytic activity for electro-oxidation of small molecules. *Advanced Materials*, *24*, 2326–2331.

CHAPTER 6

Proton Exchange Membrane Fuel Cell Electrodes From One-Dimensional Nanostructures

Shangfeng Du
School of Chemical Engineering, University of Birmingham, Birmingham, United Kingdom

The final goal for the development of one-dimensional (1D) nanostructures for proton exchange membrane fuel cells (PEMFCs) is to apply them into devices to achieve enhanced power performance and durability in real-context option. With the considerable progress achieved for 1D nanostructured catalysts themselves, there has been a growing interest in examining 1D Pt nanostructures within operating PEMFCs. Besides the conventional electrode structure fabricated from catalyst ink by mixing catalysts with electrolyte ionomer, novel electrode structures, in particular with thin film catalyst layer from aligned 1D nanostructure have also been successfully incorporated into fuel cell devices. The transition of 1D nanostructured catalysts from ex situ electrochemical measurement into operating PEMFCs is encouraging and further highlights the potential of 1D catalysts as plausible alternatives for traditional zero-dimensional (0D) nanoparticle catalysts. In this chapter, we will provide a detailed discussion of the recent advances in 1D catalysts incorporated within functional PEMFC electrodes.

6.1 CONVENTIONAL ELECTRODE STRUCTURE FROM 1D CATALYSTS

The commonly used electrode fabrication technique for PEMFCs is via catalyst ink. Usually, 0D catalyst nanoparticles mix with electrolyte ionomer and organic solvent to make catalyst ink, which is then applied (e.g., painting, printing, screening, etc.) to the surface of gas diffusion layers (GDLs) or polymer electrolyte membrane (PEM) to fabricate catalyst electrodes. Both approaches are named as gas diffusion electrodes (GDEs) and catalyst-coated membrane (CCM),

One-Dimensional Nanostructures for PEM Fuel Cell Applications. DOI: http://dx.doi.org/10.1016/B978-0-12-811112-3.00006-6

respectively (Zhang, 2008). Carbon-supported Pt nanowires (NWs) have been fabricated into PEMFC cathodes using this approach and tested in single cells (Lee & Kim, 2013; Li et al., 2014, 2015; Sung, Chang, & Ho, 2014). A test has also been successfully conducted within a 1.5-kW PEMFC stack under collaboration between Tongji University, University of Waterloo, and General Motors (Li et al., 2015). Both power performance and durability were evaluated (Fig. 6.1). Although Pt NWs had a larger diameter (c.4 nm) compared with the conventional Pt nanoparticles (c.2−4 nm), a similar power performance was still achieved benefiting from the unique surface catalytic activity of the Pt NW and the reduced mass transfer resistance in electrodes with enhanced porosity. The characterization for the catalysts before and after the durability test further indicated the enhanced stability of Pt NWs over Pt/C. However, the large porosity also resulted in a thicker catalyst layer and a looser electrode structure in PEMFCs. Although excellent stability is observed for Pt NW catalysts themselves, the improvement of their electrode durability is badly limited by the poor electrode structure fabricated from them. After a 420-h dynamic drive cycle durability test, PEMFC stacks from carbon-supported Pt NWs exhibited a performance degradation rate of 14.4%, as compared with 17.9% for that from Pt/C nanoparticle-based cathodes. The authors ascribed the performance loss to the degradation of the Pt/C used in the anode. However, considering the much easier hydrogen oxidation reaction and the high catalyst loading of 0.2 mg_{Pt} cm^{-2} used in the anode, this performance loss might be mainly due to the degradation of the cathode structure such as the decay of electrolyte ionomer, their contact with catalysts, and even carbon support corrosion, as pointed out by Holdcroft in his review on fuel cell catalyst layers (Holdcroft, 2014).

Comparing with 0D nanoparticles, 1D nanostructures with the anisotropic morphology are usually very difficult to fabricate into fuel cell electrodes by a conventional process as used for Pt/C nanoparticles. As an example, in the ongoing US DoE (Department of Energy) project in Extended Surface Electrocatalyst Development, Pt NWs showing an outstanding intrinsic catalytic activity of 1800 mA/mg_{Pt} (275 mA/mg_{Pt} for Pt/C catalysts) only demonstrated a low activity of 165 mA/mg_{Pt} (216 mA/mg_{Pt} for Pt/C electrodes) in the PEMFC cathode (Pivovar, 2015). 1D nanostructures address the major degradation mechanisms of Pt nanoparticles in fuel cells such as nanoparticle aggregation,

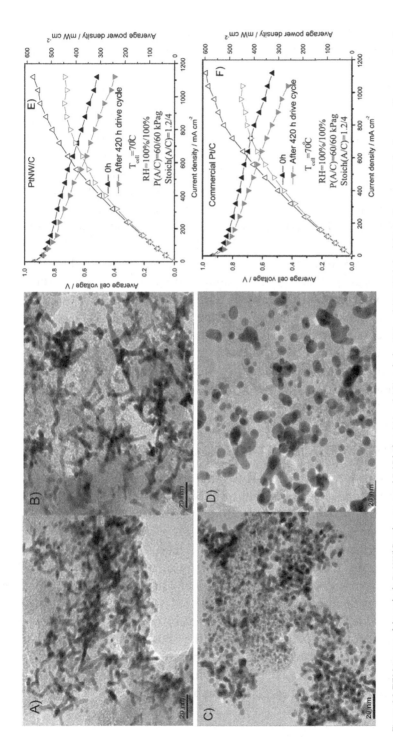

Figure 6.1 TEM images of the cathode Pt NW/C and commercial Pt/C before (A) and (C) and after (B) and (D) 420 h durability testing, respectively. Average cell power as a function of the operation time, I−V and I−P curves of (E) Pt NW/C and (F) commercial Pt/C. TEM, transmission electron microscopy; NW, nanowire. Adapted from Li, B., Higgins, D.C., Xiao, Q.F., Yang, D.J., Zhng, C.M., Cai, M., … Ma, J.X. (2015). The durability of carbon supported Pt nanowire as novel cathode catalyst for a 1.5 kW PEMFC stack. Applied Catalysis B: Environmental, 162, 133−140, with permission from Elsevier.

surface atom dissolution, and Oswald ripening. However, if the carbon support is still used, then carbon corrosion in operation and the detachment of catalysts from carbon support cannot be avoided, as discussed earlier in the 1.5-kW PEMFC stack. Furthermore, besides the issues with catalysts themselves, in fuel cell operation, the decay of the adhesion between electrolyte ionomer and catalyst nanostructures is also a major reason causing the degradation of electrodes (Holdcroft, 2014). The anisotropic morphology of 1D nanostructured catalysts, compared with conventional spherical nanoparticles, leads to a different adhesion behavior with electrolyte ionomer in the practical electrode environment. A research has recently been undertaken by adding electrospun Pt NWs to Pt/C nanoparticle cathodes (Sung et al., 2014). It is found that a small ratio of Pt NWs, e.g., up to 20 wt% improved the electrode performance, but a higher ratio resulted in a lower power performance. The research also showed a lower optimal Nafion ionomer loading in the electrode indicating a different electrolyte ionomer dispersing behavior over the conventional Pt/C electrode. Hence, novel nanostructured electrodes different from the conventional concept used for Pt/C nanoparticles have been explored for 1D nanostructures, in particular the one without using carbon support and electrolyte ionomer, as very recently reviewed by Lu, Du, and Steinberger-Wilckens (2016b).

6.2 NANOSTRUCTURED CATALYST ELECTRODES

One important progress in PEMFC electrodes in recent decades is the concept of the nanostructured thin film (NSTF) catalyst layer from 1D nanostructures. NSTF catalyst electrode, pursued by Debe et al. from 3M, has attracted many efforts. The progress of this concept has been thoroughly reviewed recently in his work (Debe, 2012a; 2012b). In the NSTF electrode, the catalyst layer consists of a monolayer array of perylene whiskers (1 μm tall, 30 nm × 55 nm in cross-section) with a surface coated 20 nm polycrystalline PtM (M = Fe, Ni, Co, Mn, etc.) alloy film, which is achieved by a decal substrate transfer approach. Scanning electron microscopy images of the alloy catalyst sputter coating on the microstructured substrate-supported whiskers before transferring are shown in Fig. 6.2. This thin film catalyst layer with a vertically aligned porous structure enables a much higher catalyst

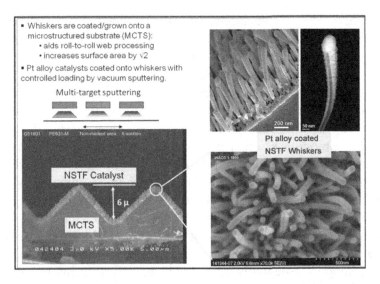

Figure 6.2 Pt alloy catalyst sputter coating on the microstructured substrate-supported whiskers for the fabrication of NSTF catalyst electrodes. NSTF, nanostructured thin film. Adapted from Debe, M.K. (2012b). Nanostructured thin film electrocatalysts for PEM fuel cells—a tutorial on the fundamental characteristics and practical properties of NSTF catalysts. *Tutorials on Electrocatalysis in Low Temperature Fuel Cells, 45,* 47—68 and Cullen, D.A., Lopez-Haro, M., Bayle-Guillemaud, P., Guetaz, L., Debe, M.K., & Steinbach, A.J. (2015). Linking morphology with activity through the lifetime of pretreated PtNi nanostructured thin film catalysts. *Journal of Materials Chemistry A, 3,* 11660—11667, with permission from ECS Publications and the Royal Society Chemistry.

utilization ratio in the electrode to achieve an enhanced mass activity. Furthermore, electrolyte ionomer is not necessary within this unique structure, due to (1) the resistance for the transport of protons is not very large for such a small distance (<1 μm) within the thin catalyst layer, and (2) the electrostatic interactions with the surface charge of the pore walls facilitates the proton transport. This electrostatic interaction is formed because Pt alloy covers the entire surface of support whiskers and nanosized hydrophilic pores between whiskers are typically flooded with water in operation. Despite these unique advantages, the approach is intrinsically limited by severe challenges of the water flooding and the very low electrochemically effective surface area (ECSA, only $10-15 \, \mathrm{m^2 \, g_{Pt}^{-1}}$). The flooding accelerates the dissolution of transitional metals in the alloy catalyst, resulting in excess metal cations in membrane, which thus enhances the net water transport across the membrane from anode to cathode and leads to increased flooding. Several approaches including acid treatment and annealing have been proposed recently to form a Pt-rich surface layer for catalyst whiskers to achieve better durability (Cullen et al., 2015).

Another advance in electrodes with aligned 1D catalysts was achieved by Du and colleagues (Du & Pollet, 2012; Du, 2010; Du, Millington, & Pollet, 2011; Sui et al., 2013). Nanostructured GDEs were prepared by in situ growing single-crystal Pt NWs on GDL surfaces, taking the advantages of the unique simplicity of the formic acid reduction approach at room temperature (Sun et al., 2011). The GDL was directly used as support substrate. The whole catalyst layer contains only a monolayer array of ultrathin single-crystal Pt NWs with a diameter of c. 4 nm and a length of 20–200 nm. The obtained structure can be directly used as PEMFC electrodes. Compared with the 3M's NSTF electrode, the ultrathin NW can provide a larger surface area, and a similar ECSA was achieved in membrane electrode assembly (MEA) to that of Pt/C nanoparticles (Du et al., 2014; Lu, Du, & Steinberger-Wilckens, 2015). Furthermore, the successful removal of catalyst support potentially contributed to a further improvement of the electrode durability. However, due to the uneven surface of the GDL and the hydrophobic surface property of Pt NWs, electrolyte ionomer is still required for the Pt NW catalyst layer to facilitate the formation of proton transport network within the whole electrode and to achieve a good contact with PEM (Du et al., 2011).

It was also found that the distribution of Pt NWs on the GDL had a large impact on the final electrode performance. An optimal growing temperature for NWs can partially balance the contact between the aqueous reaction solution and the super hydrophobic GDL surface, improving the distribution of Pt NWs on GDL surface and enabling a better electrode structure (Lu et al., 2015). A double mass activity and three times higher surface activity were observed over the TKK Pt/C cathode (45.9 wt% Pt/C, TEC10E50E) in MEA. The accelerated stress test also confirmed a better durability of Pt NW electrodes with 48% loss in the ECSA compared with 67% loss in Pt/C nanoparticle electrode. In order to achieve a uniform distribution on GDL surface and further reduce the diameter of Pt NWs for a larger surface area and enhanced electrode performance, N-doping was introduced to GDL surface by active screen plasma nitriding (ASPN) before Pt NW growth (Du et al., 2014; Lin, Lu, Du, Li, & Dong, 2016). Images of Pt NWs grown on the ASPN-treated GDL are shown in Fig. 6.3. The nitrogen doping to carbon nanospheres on the GDL surface introduced in ASPN confines the Pt atoms in reaction to form tiny nuclei and finally produced ultrathinner Pt NWs with a diameter of only c. 3 nm, offering

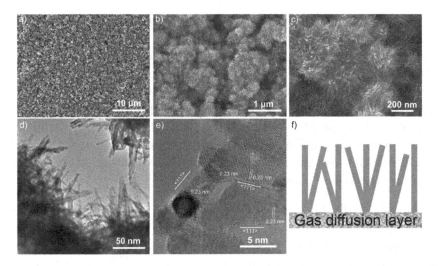

Figure 6.3 Images of Pt NWs grown on the ASPN-treated GDL. (A−C) SEM images of a 3D nanostructured catalyst layer with Pt NW arrays in situ grown on the treated GDL, at three different magnifications. The GDL size is 5 cm². (D and E) TEM and HR-TEM images of Pt NWs. (E) shows a HR-TEM image indicating the single-crystal nanowires with the growth direction along the <111> axis. (F) Schematic illustration of the nanowire GDE. ASPN, active screen plasma nitriding; GDL, gas diffusion layers; SEM, scanning electron microscopy; NW, nanowire; TEM, transmission electron microscopy; HR-TEM, high-resolution transmission electron microscopy; GDE, gas diffusion electrode. Adapted from Du, S.F., Lin, K.J., Malladi, S.K., Lu, Y.X., Sun, S. H., Xu, Q.... Dong, H.S. (2014). Plasma nitriding induced growth of Pt-nanowire arrays as high performance electrocatalysts for fuel cells. *Scientific Reports, 4,* 6439.

a larger ECSA for a better catalytic activity. Furthermore, the functional groups introduced by ASPN on the GDL surface, e.g., C−N, C = N, and −OH, facilitates the contact between the substrate GDL surface and the reaction solution to form a much uniform distribution of NWs. A larger Pt oxide reduction potential was observed on the cathode cyclic voltammogram, further confirming the weakening of the bond between oxygen-containing species and the surface of Pt NWs, which also contributed to a better oxygen reduction reaction performance. The testing of a Pt NW cathode with only half catalyst loading showed an even higher power performance than that of Pt/C nanoparticles. Inspired by the work from Xia's group (Lim et al., 2009), Pd nanoparticles were also introduced onto GDL surface as seeds to direct the growth of Pt NWs (Lu, Du, & Steinberger-Wilckens, 2016a). This also partially improved the distribution of 1D catalysts on GDL surface, achieving a higher power performance at 0.6 V. However, the introduced nanoseeds also led to the formation of dendrite-like Pt nanostructures, resulting in a less catalytic activity and poorer durability compared with the pure Pt NW electrode.

With the same formic acid approach, by growing Pt NWs on carbon nanosphere-coated Nafion membrane or PTFE surface followed by a transfer step to Nafion membrane surface, Sui et al. (Su et al., 2014; Wei, Su, Sui, He, & Du, 2015; Yao et al., 2013) fabricated this type of Pt NW electrodes which are similar to CCM and decal methods used for conventional Pt/C electrodes, respectively. In this case, a better contact between the catalyst and PEM can be achieved. Preliminary tests confirmed a better performance compared with Pt/C electrodes and also the importance of Pt NW distribution in controlling the electrode structure.

However, till date, within NW electrodes, the Pt NW is still the only 1D nanostructure that has been tested in PEMFCs. The development of electrodes with Pt alloy NW arrays can potentially enhance the electrode power performance. A long-term stability testing, in particular within PEMFC stacks, is also urgently required to confirm the real potential of this advanced approach.

REFERENCES

Cullen, D. A., Lopez-Haro, M., Bayle-Guillemaud, P., Guetaz, L., Debe, M. K., & Steinbach, A. J. (2015). Linking morphology with activity through the lifetime of pretreated PtNi nanostructured thin film catalysts. *Journal of Materials Chemistry A, 3*, 11660–11667.

Debe, M. K. (2012a). Electrocatalyst approaches and challenges for automotive fuel cells. *Nature, 486*, 43–51.

Debe, M. K. (2012b). Nanostructured thin film electrocatalysts for PEM fuel cells—a tutorial on the fundamental characteristics and practical properties of NSTF catalysts. *Tutorials on Electrocatalysis in Low Temperature Fuel Cells, 45*, 47–68.

Du, S. F. (2010). A facile route for polymer electrolyte membrane fuel cell electrodes with in situ grown Pt nanowires. *Journal of Power Sources, 195*, 289–292.

Du, S. F., & Pollet, B. G. (2012). Catalyst loading for Pt-nanowire thin film electrodes in PEFCs. *International Journal of Hydrogen Energy, 37*, 17892–17898.

Du, S. F., Millington, B., & Pollet, B. G. (2011). The effect of Nafion ionomer loading coated on gas diffusion electrodes with in-situ grown Pt nanowires and their durability in proton exchange membrane fuel cells. *International Journal of Hydrogen Energy, 36*, 4386–4393.

Du, S. F., Lin, K. J., Malladi, S. K., Lu, Y. X., Sun, S. H., Xu, Q., ... Dong, H. S. (2014). Plasma nitriding induced growth of Pt-nanowire arrays as high performance electrocatalysts for fuel cells. *Scientific Reports, 4*, 6439.

Holdcroft, S. (2014). Fuel cell catalyst layers: a polymer science perspective. *Chemistry of Materials, 26*, 381–393.

Lee, W. H., & Kim, H. (2013). Electrocatalytic activity and durability study of carbon supported Pt nanodendrites in polymer electrolyte membrane fuel cells. *International Journal of Hydrogen Energy, 38*, 7126–7132.

Li, B., Higgins, D. C., Xiao, Q. F., Yang, D. J., Zhng, C. M., Cai, M., . . . Ma, J. X. (2015). The durability of carbon supported Pt nanowire as novel cathode catalyst for a 1.5 kW PEMFC stack. *Applied Catalysis B: Environmental, 162*, 133–140.

Li, B., Yan, Z., Higgins, D. C., Yang, D., Chen, Z., & Ma, J. (2014). Carbon-supported Pt nanowire as novel cathode catalysts for proton exchange membrane fuel cells. *Journal of Power Sources, 262*, 488–493.

Lim, B., Jiang, M. J., Camargo, P. H. C., Cho, E. C., Tao, J., Lu, X. M., . . . Xia, Y. N. (2009). Pd-Pt bimetallic nanodendrites with high activity for oxygen reduction. *Science, 324*, 1302–1305.

Lin, K. J., Lu, Y. X., Du, S. F., Li, X. Y., & Dong, H. S. (2016). The effect of active screen plasma treatment conditions on the growth and performance of Pt nanowire catalyst layer in DMFCs. *International Journal of Hydrogen Energy, 41*, 7622–7630.

Lu, Y. X., Du, S. F., & Steinberger-Wilckens, R. (2015). Temperature-controlled growth of single-crystal Pt nanowire arrays for high performance catalyst electrodes in polymer electrolyte fuel cells. *Applied Catalysis B: Environmental, 164*, 389–395.

Lu, Y. X., Du, S. F., & Steinberger-Wilckens, R. (2016b). One-dimensional nanostructured electrocatalysts for polymer electrolyte membrane fuel cells—a review. *Applied Catalysis B: Environmental, 199*, 292–314.

Lu, Y., Du, S., & Steinberger-Wilckens, R. (2016a). Three-dimensional catalyst electrodes based on PtPd nanodendrites for oxygen reduction reaction in PEFC applications. *Applied Catalysis B: Environmental, 187*, 108–114.

Pivovar, B. Extended, continuous Pt nanostructures in thick, dispersed electrodes. *National Renewable Energy Laboratory. U.S. DOE 2015 Annual Merit Review Proceedings*, Arlington Virginia, USA, 2015.

Su, K., Sui, S., Yao, X., Wei, Z., Zhang, J., & Du, S. (2014). Controlling Pt loading and carbon matrix thickness for a high performance Pt-nanowire catalyst layer in PEMFCs. *International Journal of Hydrogen Energy, 39*, 3397–3403.

Sui, S., Zhuo, X. L., Su, K. H., Yao, X. Y., Zhang, J. L., Du, S. F., & Kendall, K. (2013). In situ grown nanoscale platinum on carbon powder as catalyst layer in proton exchange membrane fuel cells (PEMFCs). *Journal of Energy Chemistry, 22*, 477–483.

Sun, S., Zhang, G., Geng, D., Chen, Y., Li, R., Cai, M., & Sun, X. (2011). A highly durable platinum nanocatalyst for proton exchange membrane fuel cells: multiarmed starlike nanowire single crystal. *Angewandte Chemie, 123*, 442–446.

Sung, M. T., Chang, M. H., & Ho, M. H. (2014). Investigation of cathode electrocatalysts composed of electrospun Pt nanowires and Pt/C for proton exchange membrane fuel cells. *Journal of Power Sources, 249*, 320–326.

Wei, Z., Su, K., Sui, S., He, A., & Du, S. (2015). High performance polymer electrolyte membrane fuel cells (PEMFCs) with gradient Pt nanowire cathodes prepared by decal transfer method. *International Journal of Hydrogen Energy, 40*, 3068–3074.

Yao, X. Y., Su, K. H., Sui, S., Mao, L. W., He, A., Zhang, J. L., & Du, S. F. (2013). A novel catalyst layer with carbon matrix for Pt nanowire growth in proton exchange membrane fuel cells (PEMFCs). *International Journal of Hydrogen Energy, 38*, 12374–12378.

Zhang, J. (2008). *PEM fuel cell electrocatalysts and catalyst layers: fundamentals and applications.* London: Springer.

CHAPTER 7

Summary and Perspective

Shangfeng Du

School of Chemical Engineering, University of Birmingham, Birmingham, United Kingdom

Ever since the use of SBA-15 template nanoreactor in 2003 for synthesizing the nanowire network for methanol oxidation reaction application, an impressive progress has been achieved in the design, preparation, and evaluation of one-dimensional (1D) nanostructured materials as active and durable electrocatalysts for potential applications in proton exchange membrane fuel cells (PEMFCs). Owing to the inherent properties of unique anisotropic morphology and special surface features, 1D nanostructured electrocatalysts have shown the potential to resolve many issues associated with their zero-dimensional (0D) counterparts and are regarded as the promising candidates for the replacement of contemporary 0D catalysts.

Among all 1D catalysts, Pt-based nanocomposites are still the most practical catalysts due to their good electrocatalytic activity and durability. However, the rational design of self-supported, highly active, and long-time lasting catalysts is still a challenge, especially when taking into account the complex reaction mechanism and harsh fuel cell operational conditions. Although great efforts have been devoted to fabricate non-Pt or even nonprecious metal-based catalysts to reduce the production cost, they usually suffer from dissolution in acidic electrolyte and most of them can only perform in alkaline medium. At present, numerous works are focusing on fabricating advanced catalysts possessing combined merits of 1D morphology, multicomposition, and novel structures like core-shell, porous, hollow, and ultrathin shape. Nonetheless it seems there is still no clear understanding about the function of these factors at molecule reaction level. We also need to pay attention to that the majority of the evaluation processes for 1D nanostructures are only based on ex-situ half-cell measurement in liquid electrolyte, and only some Pt-based nanowires have been really tested in fuel cells, which is crucial for practical applications.

One-Dimensional Nanostructures for PEM Fuel Cell Applications. DOI: http://dx.doi.org/10.1016/B978-0-12-811112-3.00007-8

In light of these shortfalls, future efforts need to focus on the key aspects for the practical development of 1D catalyst for functional PEMFC devices:

1. An emphasis needs to be placed on the validation of catalyst performance in functional PEMFC devices as catalyst will experience a much different and crucial environment in real fuel cell operating conditions. Most 1D catalyst approaches developed only stay at the "test tube" level and do not really work in fuel cells, resulting in an increasing gap between the pure material research and the high-performance fuel cell device. If 1D catalysts with which excellent catalytic activities have been confirmed can be fully approved within fuel cells, the commercial potential with these catalyst systems can then be clarified.

2. More in-depth theoretical and experimental studies are demanded to understand the structure–property correlations in electrocatalysis, in particular in functional fuel cell electrodes. The anisotropic morphology of 1D nanostructures requires further fundamental understanding to develop new approaches to fabricate them into fuel cell electrodes, in particular their adhesion with water and/or electrolyte ionomer within functional electrode environment to develop novel electrode concept and structure. Without this, it will be impossible to really make 1D catalysts work by just using conventional processes for Pt/C nanoparticle electrodes.

3. Nanostructured catalyst layer with aligned could be a major direction for future development for these catalyst systems. Nanostructured thin film (NSTF) catalyst electrode, Pt nanowire array electrodes, and other nanostructured membrane electrode assemblies have confirmed the advantages of this unique structure. With the NSTF the severe challenges of water flooding and low Pt surface area need to be further addressed. While the approach for electrodes with the monolayer array of Pt nanowires could be a promising technology, this can only happen after the long-term stability and water management issues have been confirmed. Based on the much enhanced mass transfer performance and catalyst utilization ratio, if the enhanced effects of alloy and hybrid nanostructures can be brought into the structure, it could offer a bright horizon for this catalyst system.

4. An on-going pursuit should be performed to develop facile, green, and scale-up synthesis processes for 1D catalysts to realize high-yield catalyst production, in particular for approaches leading to reduced precious metal loading, i.e., 1D nonprecious group metal catalysts which could play a special role in meeting the requirement for fuel cell commercialization.

5. Continuous efforts should be made to understand fundamental catalytic mechanisms of 1D nanostructures toward oxygen reduction reaction and hydrogen carbon oxidation, including the appropriate adsorption for reactant species and favorable electron transfer pathway to optimize the geometry, composition, and structure for a further improvement of the catalytic activity and durability, especially for the C−C cleavage in ethanol oxidation reaction to meet the increasing commercial potential.

It is believed that with the resolving of these challenges, 1D catalysts with novel structural motifs, diverse advantages, multifunctional performance, and effective cost will play a significant role in achieving high-performance and robust PEMFCs.

INDEX

Note: Page numbers followed by "*f*" refer to figures.

Printed in the United States
By Bookmasters